Cambridge Imperial and Post-Colonial Studies Series

General Editors: **Megan Vaughan**, Kings' College, Cambridge and **Richard Drayton**, King's College London
This informative series covers the broad span of modern imperial history while also exploring the recent developments in former colonial states where residues of empire can still be found. The books provide in-depth examinations of empires as competing and complementary power structures encouraging the reader to reconsider their understanding of international and world history during recent centuries.

Titles include:

Peter F. Bang and C. A. Bayly (*editors*)
TRIBUTARY EMPIRES IN GLOBAL HISTORY

Gregory A. Barton
INFORMAL EMPIRE AND THE RISE OF ONE WORLD CULTURE

James Beattie
EMPIRE AND ENVIRONMENTAL ANXIETY, 1800–1920
Health, Aesthetics and Conservation in South Asia and Australasia

Rachel Berger
AYURVEDA MADE MODERN
Political Histories of Indigenous Medicine in North India, 1900–1955

Robert J. Blyth
The Empire of the Raj
Eastern Africa and the Middle East, 1858–1947

Rachel Bright
CHINESE LABOUR IN SOUTH AFRICA, 1902–10
Race, Violence, and Global Spectacle

Larry Butler and Sarah Stockwell
THE WIND OF CHANGE
Harold Macmillan and British Decolonization

Kit Candlin
THE LAST CARIBBEAN FRONTIER, 1795–1815

Hilary M. Carey (*editor*)
EMPIRES OF RELIGION

Nandini Chatterjee
THE MAKING OF INDIAN SECULARISM
Empire, Law and Christianity, 1830–1960

Esme Cleall
MISSIONARY DISCOURSE
Negotiating Difference in the British Empire, c.1840–95

Michael S. Dodson
Orientalism, Empire and National Culture
India, 1770–1880

Jost Dülffer and Marc Frey (*editors*)
ELITES AND DECOLONIZATION IN THE TWENTIETH CENTURY

Bronwen Everill
ABOLITION AND EMPIRE IN SIERRA LEONE AND LIBERIA

Ulrike Hillemann
ASIAN EMPIRE AND BRITISH KNOWLEDGE
China and the Networks of British Imperial Expansion

B. D. Hopkins
THE MAKING OF MODERN AFGHANISTAN

Iftekhar Iqbal
THE BENGAL DELTA
Ecology, State and Social Change, 1843–1943

Brian Ireland
THE US MILITARY IN HAWAI'I
Colonialism, Memory and Resistance

Robin Jeffrey
POLITICS, WOMEN AND WELL-BEING
How Kerala became a 'Model'

Gerold Krozewski
MONEY AND THE END OF EMPIRE
British International Economic Policy and the Colonies, 1947–58

Javed Majeed
Autobiography, Travel and Post-National Identity

Francine McKenzie
Redefining the Bonds of Commonwealth 1939–1948
The Politics of Preference

Gabriel Paquette
ENLIGHTENMENT, GOVERNANCE AND REFORM IN SPAIN AND ITS EMPIRE 1759–1808

Sandhya L. Polu
PERCEPTION OF RISK
Policy-Making On Infectious Disease in India 1892–1940

Jennifer Regan-Lefebvre
IRISH AND INDIAN
The Cosmopolitan Politics of Alfred Webb

Sophus Reinert, Pernille Røge
THE POLITICAL ECONOMY OF EMPIRE IN THE EARLY MODERN WORLD

Ricardo Roque
Headhunting and Colonialism
Anthropology and the Circulation of Human Skulls in the Portuguese Empire, 1870–1930

Jonathan Saha
LAW, DISORDER AND THE COLONIAL STATE
Corruption in Burma c.1900

Michael Silvestri
IRELAND AND INDIA
Nationalism, Empire and Memory

John Singleton and Paul Robertson
ECONOMIC RELATIONS BETWEEN BRITAIN AND AUSTRALASIA 1945–1970

Miguel Suárez Bosa
ATLANTIC PORTS AND THE FIRST GLOBALISATION C. 1850–1930

Julia Tischler
LIGHT AND POWER FOR A MULTIRACIAL NATION
The Kariba Dam Scheme in the Central African Federation

Aparna Vaidik
IMPERIAL ANDAMANS
Colonial Encounter and Island History

Erica Wald
VICE IN THE BARRACKS
Medicine, the Military and the Making of Colonial India, 1780–1868

Jon E. Wilson
THE DOMINATION OF STRANGERS
Modern Governance in Eastern India, 1780–1835

Cambridge Imperial and Post-Colonial Studies Series
Series Standing Order ISBN 978–0–333–91908–8 (Hardback)
978–0–333–91909–5 (Paperback)
(outside North America only)

You can receive future titles in this series as they are published by placing a standing order. Please coxntact your bookseller or, in case of difficulty, write to us at the address below with your name and address, the title of the series and the ISBN quoted above.

Customer Services Department, Macmillan Distribution Ltd, Houndmills, Basingstoke, Hampshire RG21 6XS, England

---

Vice in the Barracks

Medicine, the Military and the Making of Colonial India, 1780–1868

Erica Wald
Lecturer, Goldsmiths, University of London, UK

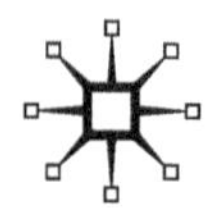

First published 2014 by
PALGRAVE MACMILLAN

Palgrave Macmillan in the UK is an imprint of Macmillan Publishers Limited, registered in England, company number 785998, of Houndmills, Basingstoke, Hampshire RG21 6XS.

Palgrave Macmillan in the US is a division of St Martin's Press LLC, 175 Fifth Avenue, New York, NY 10010.

Palgrave Macmillan is the global academic imprint of the above companies and has companies and representatives throughout the world.

Palgrave® and Macmillan® are registered trademarks in the United States, the United Kingdom, Europe and other countries.

ISBN 978–1–137–27098–6

This book is printed on paper suitable for recycling and made from fully managed and sustained forest sources. Logging, pulping and manufacturing processes are expected to conform to the environmental regulations of the country of origin.

A catalogue record for this book is available from the British Library.

A catalog record for this book is available from the Library of Congress.

Typeset by MPS Limited, Chennai, India.

Contents

List of Illustrations vii

Acknowledgements ix

Note on Transliteration, Currency and Military Ranks xi

Map xii

Introduction **1**

Unpicking the Contagious Diseases Acts 5

Approaches to the European soldier 7

(Re)Shaping Indian health and society 10

Organisation and structure 13

1 **The East India Company, the Army and Indian Society** **16**

The East India Company and its army 19

Begums and Bibis 24

The re-construction of the 'prostitute' 37

Conclusion 45

2 **Regulating the Body: Experiments in Venereal Disease Control, 1797–1831** **48**

Medical conceptions of venereal disease 51

Early experimentation with lock hospitals and regulation 54

Balancing the budget: the costs of regulation 65

'Martyrs to the effects of their licentiousness': morality and disease 69

Excuses, solutions and the production of racial and cultural stereotypes 72

Conclusion 82

3 **Medicine and Disease in the 'Age of Reform'** **84**

Surgeons and administrators in the Age of 'Reform' 86

Essays, societies and journals 94

The 1831 Bengal Medical Board circular on venereal disease 105

Journals and venereal disease 110

Conclusion 113

4 The Body of the Soldier and Space of the Cantonment **118**
Intemperance and the soldier 120
Military and medical descriptions of the European soldier 122
Canteen and cantonment: medical theories and proposals for military spaces 125
Ordering the cantonment: military and government regulations 131
Disorderly European women 140
Courts martial and punishment 146
Disgraceful and unbecoming conduct 153
Conclusion 155

5 'Unofficial' Responses to Lock Hospital Closure, 1835–1868 **157**
Responses to the closure of lock hospitals in the 1830s 160
The dispensary and charity hospital 171
Working around the abolition 176
Wars and sanitary commissions 179
Conclusion 189

Conclusion **190**

Appendix 1 196

Appendix 2 198

Notes 202

Glossary 243

Bibliography 244

Index 264

List of Illustrations

Map

Some of the larger cantonments where European regiments were stationed in British India xii

Figures

5.1 Prostitute's registration ticket 186

5.2 Ground plan of Bareilly lock hospital 188

Tables

1.1 Returns of the number of females who have been married out of the Lower Orphans [sic] School, 1 January 1800 to 31 December 1818 27

1.2 Certificates of descent and education establishment 35

2.1 Venereal disease admissions among European troops in Madras, 1802–1835 59

2.2 Monthly cost of maintaining the Kaira lock hospital 63

2.3 Venereal disease admissions at Kaira, 1 June to 30 November 1812 64

2.4 Lock hospital expenses for the months of April–June 1808 66

2.5 Lock hospital admissions, 1 January to 31 December 1808 67

2.6 Number of cases of venereal disease in European Corps, 1802–1808 67

3.1 [Extract of selected questions] relative to the nature and treatment of the venereal disease in India 107

4.1 European women who lead a dissolute life in Madras, as also of those left without protection 144

4.2 Drunkenness and punishment in Madras, 1847 148

4.3 Extract from the Character Book of the 1st Company, 1st Battalion Artillery, Agra, 28 February 1845 150

5.1 General abstract of venereal diseases in His Majesty's regiments in Bengal – from 21 December 1825 to 31 December 1831 163

5.2 Comparative state of venereal diseases treated in Her Majesty's regiments in the presidencies of Bengal and Madras from 1 January 1830 to 30 June 1838 165

5.3 Number and percentage of admissions from Syphilis Primitiva, Gonorrhoea and Hernia Humoralis among the European soldiery at Trichinopoly and Bellary from 1840 to 1848, inclusive 175

5.4 Names of common prostitutes residing in the cantonment of Cannanore, 17 November 1848 177

5.5 Diet tables for patients in lock hospitals 187

Plates

1 A nautch at Hindu Rao's House © The British Library Board Add.Or.4684 115

2 Tom Raw gets introduced to his colonel 116

3 Qui Hi pays a nocturnal visit to Dungaree 117

Acknowledgements

This book is the culmination of a long journey of research and writing that would not have been possible without many wonderful guides and companions that I met along the way. This project began as a PhD thesis at Cambridge, where I could not have wished for a better supervisor than Joya Chatterji. Her encouragement, guidance and intellectual energy throughout have been immeasurable. It was Professor Dominic Lieven, under whose tutelage I completed the MSc, whose giddy enthusiasm for the study of empires inspired me to embark on the PhD. Chai has been unflaggingly supportive over the years.

Interactions with colleagues at Cambridge, the London School of Economics and Goldsmiths have fuelled my intellectual development. For countless cups of coffee, as well as conversation and careful reading, I owe a debt to Uditi Sen, Aditya Sarkar, Onni Gust, Jonathan Saha and Rachel Johnson, all members of the Stoke Newington academic mob, which has now (sadly) been scattered to the four corners. Tanya Harmer has been a close friend and careful reader throughout. Colleagues and scholars too numerous to count have provided me with helpful insights and encouragement at various seminars and conferences, as well as over cups of chai at the archives, which have helped me to think differently about the work and refine its focus. In particular, I would like to thank Douglas Peers, Philippa Levine, Sanjoy Bhattacharya, David Arnold, Harald Fischer-Tine, Chandar Sundaram, Taylor Sherman, Rachel Berger, Debjani Das, Durba Mitra, Yasmin Khan, Paromita Das Gupta, Riyad Koya, Gavin Rand and Simin Patel. Thanks to Eleanor Newbigin, Laura Ishiguro, Rohit De, Leigh Denault and Mitra Sharafi, who have read drafts of the work and provided useful comments. In addition, the advice of my two examiners, Professors Mark Harrison and David Washbrook, helped steer the final course of the work.

I have had many enriching adventures at various archives across India and Great Britain. The librarians and archivists at these sites have been ever-friendly and helpful in locating records. In particular, I owe a great debt to the following archivists and their teams: the reference enquiries and issue desk staff at the Asia and African Studies reading room at the British Library; Jaya Ravindran and her team at the National Archives in Delhi; PD Thombre, SR Duduskar, and the rest of the team at the Maharashtra State Archives; the Tamil Nadu State Archives; the

West Bengal State Archives; the Victoria Memorial Library in Kolkata; the Asiatic Society libraries in Kolkata (with particular thanks to the staff at the Annex in Metcalfe Hall, whose encyclopaedic powers of recall were simply awe-inspiring), Bombay and London; the Council for World Mission Archives at SOAS; the University Library and Centre for South Asian Studies library at Cambridge; and the LSE library.

A healthy lifeline with the outside world came via many wonderful friends, including the women of my book group, with whom I have been discussing doses of fiction over drinks and dinner for some years now. I would also like to thank the many others who were always happy to meet up for coffee or a walk with Wilbur (who, of course, deserves recognition in his own right). By happy accident, a few years ago, I moved next door to the photographer and historian of photography, Val Wilmer, who helped to identify and date a number of images for me. Joanne Goatman was kind enough to welcome me into her house in Delhi and let me house-sit while I worked at the National Archives. Nadia Salvadori, who is sadly no longer with us, not only needled and adjusted my battered back, but was a wonderful listener. She is greatly missed.

A number of funding bodies and departments were very generous in their support of my work. In particular, I would like to express my thanks to Dr Anil Seal and the Cambridge Trusts; the Prince Consort and Thirwall Fund; the Royal Historical Society; the Member's History Fund; the Smuts Fund; the Eddington Fund; the Central Research Fund, University of London; the International History department of LSE; and the History department at Goldsmiths.

Thanks are due to Professors Megan Vaughan and Richard Drayton as the commissioning editors of the Cambridge Imperial and Post-colonial studies series and to Jen McCall and Holly Tyler at Palgrave Macmillan. The comments of an anonymous reader encouraged me to broaden the scope of the work. I am grateful to Palgrave Macmillan's copy-editing team for adjustments in the final stages. Any errors which remain, are, of course, mine.

I would like to dedicate this work to the memory of my grandmother, Rochelle Stenn, who was an inspiration to everyone who knew her. Her experiences made me realise the importance of the study of history. Her sharp wit and phenomenal recollection of even the smallest of details (including, for example, the price of different varieties of herring in the early 1930s) are just a few of the things that made her a treasure. My parents, Cy and Pamela and my sister Leslie, have always been supportive of my every endeavour. Asta arrived towards the end of this project, but I look forward to many future adventures together.

Finally, I would like to thank Marenco, whose support and love propelled me through this project.

Note on Transliteration, Currency and Military Ranks

The spelling of place names and individuals has been retained as it appears in the sources; where such names are dramatically different today (for example Futtyghur/Fatehgarh), the modern spelling is noted in a footnote. In a similar manner, charts and tables appear as they exist in the original records. The stress is not on the accuracy (or otherwise) of these statistics, but on how they were used and interpreted by contemporary observers.

In the 1820s, 1 rupee equalled approximately 2 shillings. 100 Current Rupees equalled approximately 86 Sicca Rupees, 3 Annas and 4 Pies.

To give readers a rough outline of military ranking, the lowest-ranking soldier was a private, or gunner. Above him (in ascending order), ranked Corporal, Sergeant, Sergeant-Major, Lieutenant, Captain, Major, Lieutenant-Colonel, Colonel, Brigadier, Major-General, Lieutenant-General and General.

Map

Map Some of the larger cantonments where European regiments were stationed in British India

Introduction

> The col'nel gives an intimation,
> Our youth to the barracks must repair,
> For all *the youngsters* are sent there;
> He goes, and soon a jovial set
> Initiates him – a cadet.
> Gallons of arrack, *lots* of beer,
> In fact, the *very best* of cheer,
> Was *here* prepar'd by way of *fete,*
> To give the *new cadet* a treat,
> And shew the youth an interlude,
> Before *the business* would conclude;
> They broke the windows, and in pairs,
> Dispatch'd both tables, shades, and chairs;
> And to confirm this midnight fun,
> Oft to the *loll bazaar* they run-
> The muse now *blushes* to disclose
> The bobbery that here arose;
> Our hero, being but a stranger,
> Knew nothing of impending danger;
> His new preceptors well could tell him
> Some Indian words, but could not spell'em;
> And thus the boy, *on recollection,*
> Turn'd linguist *without reflection...*[1]

For most of the nineteenth century, observers held the European soldier stationed in India in a particularly low regard. Even Rudyard Kipling, whose 1892 *Barrack Room Ballads* expresses sympathy for the plight of the Tommy, portrayed a routine almost identical to 'Qui Hi' in 1816.[2]

These fictional, stock portrayals grew out of those of military officers and army medical men who, from the early nineteenth century, increasingly lamented the debauched day-to-day existence of the European soldier in India.

Shortly after landing in India, a soldier would be forcibly initiated into a lifetime of drinking. This induction was not complete without both physical aggression (in the above case, the breaking of furniture) and a visit to the *lal* (or 'red') *bazaar,* an area in which the army's regimental prostitutes lived and worked and where the men would have their sexual 'needs' met. Thus, life in India for the common soldier followed a consistent routine, punctuated by heat and humidity of a sort not experienced in Europe (and something which many commentators maintained contributed to this combination of languor and disintegrating morals), tedious early morning drill and (only very occasionally) engagement with an enemy. Moreover, observers opined that the young soldier was exposed to a greater degree of 'vice' in India than was the case in England. 'Debauchery' in India was presented as a tangible force in its own right, hanging about the cantonments like a malign miasma.

This grim trinity – of drink, violence and sex – was so frequently associated with the European soldier that it is easy to believe that this was an ageless routine. Yet, to the extent that this was reflective of life in cantonments across India, it developed as the result of a number of factors, including perhaps unsurprisingly, the fiscal constraints that bore down on the East India Company as it expanded its political reach across India in the late eighteenth century. No doubt, the low opinion of soldiers held by Crown and Company officials was also not unique.[3] It suited many higher officials to look upon the European rank-and-file as *terra incognita* – they assumed a host of unflattering, often violent stereotypes based on the men's social backgrounds.[4] Perhaps more importantly, this vision of the animalistic, over-sexed brute allowed officials to ignore (or minimise the importance of) such things as the necessity of providing for their other basic requirements – including family, social and intellectual provisions as they completed their long years of service in India.[5] As a commercial outfit whose primary concern was maximising profit, the East India Company was no doubt also wary of having to support the growing numbers of native wives, mistresses and children who surrounded its expanding army. As such, the policy of encouraging or tolerating native companions began to change by the late eighteenth century. At the same time, each of these men represented a considerable expense on the part of the Company. Given the relatively expensive cost (at least compared to the Company's more numerous, but locally recruited sepoys), each European soldier was viewed as an asset to be protected. Yet, these were investments with uncertain returns. Not only did the European soldier in India experience much higher mortality rates when compared to his peers in Europe (or indeed, his Indian counterparts), but he often fell victim to a host of less fatal, but equally costly, ailments. At the same time unofficial wives were being discouraged, surgeons were growing increasingly alarmed at the rising levels of venereal disease among the European troops. Indeed, venereal diseases and their associated ills were an ever-present, looming menace for the colonial state. Syphilis and gonorrhoea continued to frustrate all army attempts at control, even after the discovery of a cure in the early twentieth century. The needs of the military, the actions of the medical corps and the demands of an expanding political power can all be read through the lens of how the East India Company state viewed, and dealt with, the preservation of one of its most important elements, the European soldier. This book examines not only the colonial state's approach to venereal disease and 'vice'-driven health risks, but the broader context that shaped such policies and the impact this had on Indian society. Attempts to manage 'vice' in the barracks contributed to how Europeans understood not just venereal disease and drink, but disease, race and Indian society more broadly. It sets out to understand how fears about rising levels of drink and disease dictated not only actions within the cantonment, those semi-permanent military stations across British India, but quickly spread beyond its permeable boundaries.

Military and medical responses to venereal disease and drink also highlight the fractious nature of the Company's rule. Through such a focus on 'vice' driven threats, the ambiguities, inconsistencies and tensions of colonial rule are brought into sharp focus. The Company state was not a monolith – its officials contested and renegotiated central, civil directives and flatly ignored others. Surgeons and commanding officers frequently flouted orders regarding venereal disease and the control of cantonment space. Moreover, while European troops were held to be essential to the Company's control over the subcontinent (as the embodiment of its military might), the relationship between medical and military officers and the soldiery was often an uncomfortable one. The quality of recruits to the army was a constant issue. Commanding officers were often ill-equipped to deal with the complexities that surrounded service in India. Most officers held strong class assumptions that meant that they viewed the soldiers as brutish, barely controllable louts. A number of observers even asserted that officers had no control over the men, who, in turn, openly despised them.[6] This book demonstrates how these discordant views of European soldiers – essential to Company control, yet mistrusted and feared – informed policies relating to their health.

In the late eighteenth and early nineteenth centuries, Company power had only been recently established. This was a tense and often fraught period for the Company. It still had to deal with waves of disturbances and revolts across its new territories. While the Company had turned an impressive profit in earlier years, in part due to its aggressive expansion, the later years of the eighteenth century saw the Company's finances in an increasingly precarious state. The army was frequently in arrears and defections were not uncommon. In 1781, Charles Smith, commanding in Madras, begged Warren Hastings, 'Our distress for want of money is become truly alarming, and we earnestly intreat [sic] your honour and will take measures for supplying us without loss of time.'[7] It is in this context that we need to view the response to vice-related health concerns. These represented not just an additional economic and manpower strain for the East India Company, but were seen as a threat to its very stability due to the sheer volume of men being treated at any given time.

The concern over rising military costs combined with spiking venereal rates paradoxically led to an encouragement of the more short-term sexual relations with bazaar prostitutes. With uncertain treatments and no way of guaranteeing a 'cure' for those infected, medical surgeons began to experiment with alternative means of controlling the venereal 'plague' by targeting the women who lived and worked in and around the cantonments in an attempt to guarantee a 'healthy' pool of women for the soldiers.

The lives of the many Indians whose lives and work interacted with the European soldiery have often been relegated to the insignificant periphery. This is a regrettable and serious omission. In official records, there is only indirect reference to these individuals. Rarely identified by name, they instead appear as statistics or afterthoughts. However, we ignore them at our peril. These unnamed middle men and women were critical intermediaries for the colonial state, performing a variety of social, commercial, military and political roles. At the same time, segments of this population, due to their association with the health of the military, were transformed into potential 'threats' to the stability of the East India Company state. As the nineteenth century progressed, Indian women who were deemed 'prostitutes' personified such security risks, as did those Indians resident in and around cantonments who were said to 'lure' soldiers to intoxicating drink. What follows is an examination of how military concerns over the health of the European soldiery developed into the responses to such 'threats'. It considers how these reactions, in turn, provoked an expansion of military-controlled (over civil) space.

Medical practices surrounding venereal disease were not simply European ideas transplanted to India. Instead, European surgeons absorbed and adapted the practices of their Indian counterparts, borrowing freely not just from the Indian pharmacopeia, but from 'traditional' Hindu and Muslim theories of medicine and the body. They conducted their own experiments – which varied from station to station. Surgeons then passed news of their findings to peers across India and the empire. Recent studies on the construction of medical and scientific knowledge in colonial India have poked holes in earlier theories that framed the spread of European medicine around the world as a one-way 'diffusion' of western scientific knowledge onto passive non-western recipients.[8] As has been widely argued, in areas like South Asia where active, dynamic medical and scientific cultures engaged with 'western' science, this model is impossible to justify.[9] This work expands on this idea, arguing that this engagement definitively shaped a number of European medical practices on venereal disease, many of which were in turn exported to Europe.

Unpicking the Contagious Diseases Acts

In many ways, this book works backwards from the well-researched Contagious Diseases Acts that operated in towns across Britain and the empire in the late nineteenth century.[10] Introduced as a pilot-scheme in 1864 and eventually implemented in 18 military towns, the Acts identified those women who worked as prostitutes in and around cantonments as the vectors of the venereal diseases that were 'stalking' military garrisons. As such, the women were required to register with the police and submit themselves to regular examinations to 'ensure' their health. If deemed 'healthy' following their internal examination, they would receive a notation on their 'ticket' to that effect; if 'diseased', they were sent to a lock hospital. Lock hospitals were punitive sites – places where women believed to be infected with venereal disease were forcibly detained and 'treated' against their will. Studies examining venereal disease control in Britain and colonial locations have largely focused on these late nineteenth century acts, although there have been some important exceptions.[11] In contrast, this book argues that we cannot properly understand the operation of these later measures of control without analysing their earlier Indian counterparts.

Similar acts were rolled out across the empire. Along with India, Contagious Diseases laws were enacted in Jamaica, Trinidad, Hong Kong, Fiji, Gibraltar, Malta, Burma, Ceylon, Australia, Malaya, the Cape and Cairo.[12] Philippa Levine has argued that the colonial Acts were not only more far-reaching, but were more coercive than their British counterparts.[13] While the British Acts permitted only a special group of 'enforcers' to apprehend women in breach of its rules, in Calcutta, for example, regular police officers were authorised to arrest any woman who had failed to register or attend her periodic examination.[14] From the 1870s until 1888, an average of 12 women were arrested daily for breach of the rules in Calcutta alone.[15] Despite the spread and scope of the regulations and the police surveillance that accompanied them, the Acts remained largely unsuccessful in achieving their goals. Not only did venereal levels remain persistently high, but women could, and did, regularly avoid the invasive regular inspections.[16] This outcome should have come as no surprise. The earlier experiments and regulations surrounding lock hospitals in India produced similarly disappointing results for the military.

Following the repeal of the Contagious Diseases Acts in Britain in 1886, campaigners shifted their attentions to the continued operation of the colonial Acts, with a particular focus on India. The relationship between western feminists and the Indian women targeted by the acts was complicated. Western feminists often embraced the imperial ethos and racial assumptions, seeing themselves as nationally, racially and morally superior to their Indian 'sisters', who they felt the need to 'save'.[17] The repeal campaigners and Indian public faced an uphill battle, as many colonial governments now assumed that the regulation of sexuality was essential to the proper functioning of the empire.[18] It is one of the aims of this work to draw out the means by which this supposition – linking sexuality and the maintenance of imperial control – lodged itself so firmly in the minds of military, medical and government officials in India.

It is often thought that the English and imperial Acts simply adopted the 'continental' or French model for regulated prostitution. However, as will be seen, they more closely reflected half a century of direct experience with the lock hospital system in India. Indeed, this book is, in part, an attempt to remedy the Eurocentric understanding of the emergence of the Acts. It argues that we cannot fully understand how the later Acts developed, or indeed how Britain viewed its empire, without looking at the precursors which spread across India from the late eighteenth century. In so doing, it shifts the focus from the British 'metropole' to numerous sites of experimentation and implementation across India in the first half of the nineteenth century. Moreover, it shows that in the case of venereal disease, there was a remarkable degree of continuity throughout the century. In this realm, very little, if anything, changed when the Crown assumed power in 1858.

This refocus also allows us to explore some of the ways in which military and medical knowledge travelled across a web of networks in the nineteenth century. These ranged from unofficial, casual discussions between European and Indian practitioners, to correspondence between cantonment surgeons and their peers across imperial sites, all the way to the formal circulars, statistics and orders produced by the colonial state. In tapping into these, we can see how information about venereal disease specifically, and India more broadly, flowed across the empire. Much like these intersecting networks, what follows owes a good deal to a variety of different fields of historical endeavour such as the history of gender, sexuality, the military, medicine and the colonial state (among others). This book draws from and weaves together a number of these threads to present a more holistic narrative.

Approaches to the European soldier

The medical and military anxieties that surrounded the European soldier in British India form the starting point for this study. While Indian soldiers, or sepoys, formed a much larger proportion of the fighting force, both Company and Crown viewed European troops as essential to their ability to retain their Indian possessions. These units, often under-strength and prone to high levels of mortality and morbidity, were nevertheless seen as the necessary 'steel core' of the army.[19] As its control spread across broader swathes of territory, so too did the conviction that a strong army was needed not just to provide the necessary stability for revenue collection, but to lend the Company credibility as a political ruler.[20]

Military-fiscalism dominated the decision-making of both the Company and Crown in India.[21] For the Company, the single largest budgetary item, even in years of 'peace', was defence spending. This is not to say that civil expenditure was insignificant. As Douglas Peers points out, the expenses of civil administration grew enormously in the early years of the nineteenth century. Between 1815 and 1823, military expenditure increased by 8 per cent, while the cost of civil administration shot up by 43 per cent.[22] And yet, despite this staggering growth, military costs continued to consume the largest proportion of Company expenses. Within this military bracket, the majority of the budget went towards the pay and maintenance of officers and soldiers (European and Indian). The most expensive troops were the European regiments. On average, a King's Regiment cost just under £51,500 to maintain each year, while a sepoy regiment (with the same number of men) cost just £24,500.[23] As the number of European troops in India grew, this expense increased accordingly.[24] While the European troops received approximately a shilling a day in pay, the cost of training, clothing and maintaining the men in India brought the sum considerably higher.[25] An 1830 Bengal estimate spelled out specific amounts; the European cavalryman cost £100 to maintain annually, while a European artilleryman cost £61. On the other hand, their Indian equivalents were markedly less expensive – at £64 and £28, respectively.[26] Meanwhile, a sepoy's pay in the 1830s (and indeed for most of the nineteenth century) was seven rupees per month. As Seema Alavi points out, the Company offered better terms than other professions (including, crucially, regular pay and some sense of security). However, this amount was still manifestly less than that paid to the European troops.[27] Representing as they did such a considerable expense, it was imperative that European soldiers, as valuable 'assets', be protected. This concern with expense was ever-present and led both Company and Crown to carefully weigh the costs of every eventuality. This meant that those 'avoidable' losses, such as those caused by venereal disease or drink, were especially resented.

The fiscal tensions that informed the Company's attitude toward its European troops lay behind its fixation on the men's health. The economic woes of the Company further compounded the problem of the expense of the troops. Until 1800, the Company's annual revenues covered its expenditure. However, after that time, the Company lapsed into frequent deficits. Such shortfalls did not result in a reduction in army strength. They did mean, however, that economies were sought wherever feasible. Illness amongst the men, especially due to preventable ailments, was one expense that the Company was desperate to avoid. This concern also clarifies the reasons behind the Company's willingness to employ systems like the lock hospital, given the assurance from surgeons that it would curb the number of men 'lost' to the Company as a result of venereal disease by targeting the (less expensive) Indian women whom, it was reasoned, gave the men these diseases. However, the concern with cost was only part (albeit a crucial one) of the equation that dictated the Company's approach to the men's health. A fear of its own European troops determined the exact form these regulations took. Simply put, the army was wary of any measure that might 'provoke' the men – who were seen as low-class miscreants – into mutinous behaviour. This image of the soldier, as little more than an uncivilised barbarian whose loutish nature was not unlike a rumbling volcano, persisted for much of the nineteenth century. Yet, officials viewed the men's 'manly', masculine 'urges' as essential to their effectiveness.[28] Military officials portrayed this as a delicate balancing act, arguing that the men had to be carefully managed and catered to for fear that, if pushed too far, they might explode and turn against their masters. Accordingly, they argued that 'healthy', controlled outlets were needed to provide for the men's perceived needs. This disquietude is reflected in both the lock hospital system and cantonment regulations, which sought to control the Indians who *surrounded* the European soldiery, rather than regulating the soldiers themselves. The particular anxieties about the best means to harness the strength of men from the 'lower orders' and shape them into efficient soldiers, combined with the uncertainties over the political and military position of the Company in India to dictate approaches to its soldiery.

The extremely punitive system of surveillance and treatment known as the *lal bazaar* and lock hospital system emerged as a result of this idea and the sexual and social 'threats' stalking the men. The system expanded rapidly as it moved from unofficial operation in the late eighteenth century to its official introduction in 1805. By 1822 lock hospitals were in operation at most major stations where European troops were based. In Bengal alone, they treated an estimated 4,000 women annually.[29] The system continued to operate officially until 1831, when due to its expense and lack of substantive results, Governor-General Lord Bentinck ordered its abolition (first in Bengal and four years later in Madras and Bombay).[30] The early 1830s saw perhaps the greatest challenge to the lock hospital system. However, even debates at this time focused around its cost (measured carefully against any reduction in the number of European soldiers admitted to hospital with venereal ailments). These disputes were critical in the so-called Age of Reform and emphasise the colonial state's most pressing concern at the time – how to reduce expenditure without sacrificing military integrity. Even after the government in Calcutta withdrew its support for the lock hospital system so favoured by most military surgeons, officers were left with more than enough loop-holes to allow them to continue, albeit unofficially, with similar measures.

What followed was an unofficial period of operation. However, this did not last long. In the wake of resurgent health concerns following the Crimean war and the 1857 uprising, surgeons and commanding officers were once again able to roll out the regulatory system across India. In 1859, London sanctioned an 'experimental' lock hospital in Bengal,[31] which was followed closely by Act XXII of 1864 (known as the Cantonment Act), which formalised the regulatory system of prostitution in and around cantonments. The Indian Contagious Diseases Act (Act XIV) followed in 1868.[32]

(Re)Shaping Indian health and society

It is well-understood that colonial expenditure on health was almost entirely limited to preserving that of Europeans and the military.[33] Indeed, there was barely any awareness, or real attempt made by the colonial state to gauge levels of disease among the wider Indian population until much later in the century. Instead, concerns over gonorrhoea and syphilis remained resolutely restricted to those Indian women who, medical and military officers felt, communicated the disease to the European soldiery. Nevertheless, this focus had an impact on later 'public' health measures. These earlier military policies and experiments surrounding venereal disease and the control of cantonment land served to 'normalise' certain practices and broaden the scope of what was considered politic and permissible.

One such impact can be seen when we compare later measures to control the bubonic plague in the late nineteenth century. Earlier studies have suggested that the invasive control measures which the state employed – house-to-house searches, the forced hospitalisation of suspected victims and the segregation of those deemed to be 'infected' – were unprecedented. David Arnold has described this as a 'new interventionism' on the part of the state.[34] The state's reaction to the epidemic was grossly disproportionate to the actual threat that the plague represented.[35] However, while the scale and scope of the plague measures was certainly unparalleled, the practices themselves were not new. Medical officials, with experience of earlier plague and epidemic control, used practices such as the restrictions on the movement of persons and quarantine.[36] These methods were then combined with the forms of intervention and bodily control developed by the lock hospital system, including house-to-house searches for infected persons, forced examination and compulsory treatment. In a similar manner, the mid-nineteenth century dispensaries and clinics often believed to herald the emergence of public health in India were frequently appropriated by military surgeons intent on ensuring the health of the troops.

A number of historians have pointed to the ways in which European medical practice in late colonial India was often at odds with the goals of the state.[37] However, the absolute centrality of the military to the East India Company state provided military surgeons and commanding officers with a receptive audience for demands regarding the control of these vice-related threats. An alliance of medical and commanding officers successfully moulded the colonial state's decision-making when it came to such 'moral' ills.

A focus on these ills and medical responses in this period provides us with a window into nineteenth century medical and military networks operating both formally and informally across India and the British empire. Venereal disease was an issue about which most European surgeons and commanding officers serving in India held definitive views. Understanding how these opinions were shaped illustrates how many of these networks functioned. Surgeons' struggles to contain the venereal menace amongst European soldiers coincided with a period that saw increasing moves towards 'formalised' practice in medicine in England. Many were quick to make use of the new professional societies emerging across India and Europe – employing them as a platform to promote their ideas about the control of venereal disease and intemperance.

The experiences of medical and military officials in treating venereal diseases influenced state and society in far-reaching ways. Despite their popularity with surgeons and commanding officers, these systems of control consistently failed in their aims. From the onset, it quickly became apparent that the women targeted by the lock hospital regulations found them objectionable. Many fled the cantonments regularly to avoid the internal examination, or simply moved their business to just outside the boundaries of the regulation's applicability.[38] European surgeons and officers, expressing their frustration, professed their inability to understand why women in the *profession* of prostitution were so offended by the examinations. Increasingly frustrated, surgeons produced excuses to explain the reasons why lock hospitals had not met their high hopes and promises. Their reports increasingly stressed a number of racial, social and sexual differences which surgeons proclaimed existed between Indians and Europeans that 'explained' this failure. This book maps the ways in which venereal disease, drink and 'immorality' proved central to the formulation of new ideas on race and sex which stressed the need for social, racial and sexual distance and control and contributed to the production of a medically driven 'morality'.

Historians have often inadvertently accepted certain definitions and categorisations without interrogating the processes through which they came to be constructed. The categorisation of various groups of women simply as 'prostitutes', is one such definition which demands to be deconstructed. Between the 1820s and 1850s, British rule aggressively re-structured Indian society and the economy to suit its own needs, definitely outlining and indeed shaping many of the supposedly 'backward' traditions of India.[39] Understandings of appropriate social mores and behaviour for women were not simply transplanted from Europe and quietly absorbed into Indian society. The reality was more complex and, in the case of the construction of 'appropriate' sexuality, grew out of countless earlier contestations – sexual, medical, military and intellectual (to name but a few). The role of conflicts between the military, medical doctors and the emerging Western-educated Indian elite (among others), with courtesans, temple dancing girls, *nautch* girls and bazaar women in this transformation has long been ignored.[40] While numerous practices were later condemned as 'immoral', this book examines how changes were wrought on different groups of women, including courtesans, *nautch* girls, temple dancing girls and those living around cantonments and bazaars as a direct result of military and medical demands.[41] Over a relatively short period of time, the colonial state fused various practices together without distinction, designating them singularly as 'prostitution'. This was part of a broader pattern of naming and defining groups and practices – which included not just Indian women deemed 'common prostitutes', but working class European soldiers. While it would be preposterous to claim that prostitution, or this class-based definition of the men, was 'created' by British colonialism, the colonial state heavily influenced the re-definition of these categories of women to include groups of women who formerly had very little to do with the stereotypical brothel inmate or cantonment prostitute and, likewise, made heavy use of this description of the men to serve specific demands for how the colonial state *should* function and the safeguards necessary for its smooth operation.[42] It is no coincidence that during the same period, the colonial state made sustained attempts to regulate sexuality as it attempted to control venereal diseases amongst the European soldiery. In exploring this transformation, this book speaks to a number of recent historical studies of gender, sexuality, family and prostitution in colonial India.[43]

Organisation and structure

This study begins with an examination of class assumptions that surrounded the European troops. It links this to the changing sexual relationships between European men and Indian women. The decline in monogamous relationships, perhaps unsurprisingly, saw new concerns over the increased levels of venereal disease seen among the European troops in the early nineteenth century. Those who practised prostitution in and around military cantonments where European troops were stationed were seen as threats – not just to the health of the men, but to the overall stability of the British imperial project. Over the course of the nineteenth century, these various categories of women, engaged as they were beyond the limits of 'acceptable' sexuality, had all come to be identified as 'prostitutes'. To address these issues, Chapter 1 begins by examining the transformation which took place in the late eighteenth and early nineteenth centuries, not only in the relationships between Indian women and European men, but in the construction of the 'prostitute' category. The colonial state began to exert pressure on these relationships in large part due to financial pressures. The chapter also demonstrates how, as the century progressed, definitions of 'wife', 'concubine' and 'prostitute' ossified. It shows how what came to be identified as 'prostitution' in India was, in fact, an amalgamation of a number of varied practices, beliefs and professions that spanned the social and economic spectrum. It demonstrates how this was linked to understandings of the European rank-and-file – unpicking issues such as the recruitment, composition, benefits, and (perhaps most importantly in the eyes of the Company) cost of the European soldier in India. What follows demonstrates how these understandings and constructions were fundamentally intertwined.

Chapter 2 then examines the early formation and operation of the lock hospital system, suggesting that the military's early experience of the system not only informed its later approaches to venereal disease control (most notably in the British and colonial Contagious Diseases Acts), but also shaped the colonial state's perception of India and the ways in which it was deemed acceptable to deal with 'threatening' individuals and spaces. This period of experimentation with lock hospitals established a paradigm for dealing with both venereal diseases and the Indian women assumed to be the vector by which they were spread, as well as establishing a framework for providing broader control over colonial disease.[44] The lock hospital, or regulatory system, created a means for monitoring the bazaar women, inspecting them regularly for any sign of infection. Chapter 2 maps out the long-term processes that led to the constitutional 'crisis' which Philippa Levine describes[45] and suggests the consequences, both procedural and social, of the colonial state's earlier attempts to control 'vice'.

Chapter 3 analyses the impact of the convergence of a number of forces in European medicine that took place in the 1830s. It examines the formalisation of medicine and science in early nineteenth century India. Viewing any set of medical practices, be it unani, ayurveda or 'western' medicine, as a 'system' is, as many historians have pointed out, inherently problematic.[46] This chapter does not seek to do so. What it does, however, is explore the role played by the medical boards, scientific and 'learned' societies, and their nascent journals in professionalising European medicine on the subcontinent and suggests that these institutions were increasingly utilised by European surgeons eager to share their 'unique' knowledge of the subcontinent whether on medical, social or political issues.[47] Not only was there a more formal structure for selecting candidates for medical positions in India and active medical boards in each of the presidencies, but the emergence of these medical and scientific societies gave European practitioners in India added confidence, as well as a forum to discuss their experiences and theories. This chapter suggests the ways in which this platform helped to solidify medical demands, resulting in the promotion of a 'common sense' on venereal disease control. In so doing, it highlights some of the ways in which information was gathered and transmitted across the empire. In the case of venereal disease, we see that surgeons and commanding officers began a more effective lobbying campaign, deftly making use of these new outlets, and exerted pressure on the state to heed their demands regarding the soldiers' health.

To address the gap that exists in our knowledge of attempts to control the soldier's behaviour – especially their fondness for drink – Chapter 4 looks at a number of aggressive measures of control enacted in and around the cantonments in the attempt to manage this risk. Moreover, it highlights how a number of reports stressed the need to separate European soldiers from the Indian 'other' to protect against these vice-driven threats and suggests the ways in which this contributed to race-based thinking. Through their constant lobbying of government and the medical and military boards, medical and commanding officers sought, and were granted, greater control not just over the bodies of those who represented such threats, but over the physical space which the soldiers travelled.

Chapter 5 returns to the issue of venereal disease control to examine the actions of surgeons in the period from the official abolition of the lock hospital system in 1831 until its re-instatement in the 1850s and

1860s. It argues that the system of surveillance and forced treatment set up during the years of its official operation was never fully dismantled. When the system fell from official favour, surgeons and commanding officers moved quickly to reproduce it, making use of such things as the newly established charitable dispensaries for their more explicitly military purposes. For surgeons and commanding officers, the system became the dominant paradigm for how to deal with venereal disease in India and despite all evidence which pointed to its complete lack of success, surgeons clung to the lock hospital 'ideal' for fear of a potential venereal explosion among the troops which would, in turn, destabilise the colonial state. Chapter 5 argues that the parliamentary commissions, ordered in the wake of the Crimean war and 1857 uprising on the health and organisation of the army in India, simply reinforced contemporary medical thinking (or pressures) in India. These commissions paved the way for the broader, official re-introduction of a regulatory system already operating sporadically across the subcontinent.

Too often, historians have accepted the designation of 'prostitute' without question. In the same way, ideas about the formation of certain social mores and 'traditions' have remained unexplored and unchallenged. This book analyses the motivations that prompted not just the imperial state, but elite groups of Indians, to re-shape certain 'moral' categories – medically, socially and legally. It suggests some of the consequences of these actions; many of which are still felt today. The interlinked fields of medical, social and imperial history have greatly expanded over the past 30 years. This book is situated at the intersection of a number of these fields in the hope that by connecting an analysis of military concerns and medical practice, as well as broader strands of gender, social and imperial history, a fuller picture will emerge. It contends that it is only by drawing connections between the various foci of historical study – whether it be the links between changing ideas of the family and military demands or the impact of missionary portrayals of Indian society on changing understandings of its women – that we can understand the altered social mores in nineteenth century India. Despite the fact that the measures endorsed by European surgeons wholly failed to control 'vice'-related ills, they remained strongly supported by the colonial state. The experiences of surgeons in fighting for the retention and expansion of the lock hospital system proved formative – not simply in dealing with venereal disease. Moreover, their demands, and the policies that resulted not only forcibly re-configured roles and traditions, but, more broadly, shaped European conceptions of India.

1
The East India Company, the Army and Indian Society

By the time S. M. Edwardes wrote *Crime in India* in 1924, the widely held belief among his contemporaries was that 'prostitution had existed in India from time immemorial' and, further, that a specific prostitute 'caste' could be clearly identified.[1] As late as the Second World War, regulations were in place that targeted the 'habitual spreader of venereal diseases' – the brothel prostitute – instead of acknowledging the male soldiers' role in transmission.[2] These sentiments had become closely interlinked. Yet until the early nineteenth century, monogamous relationships between European men and Indian women were common and prostitution in India, where it existed, was not viewed with the same venomous distaste as displayed in later years. The move away from official approval (or at the very least, toleration) of these long-term relationships towards encouraging the short-term liaisons that prostitutes provided, was closely linked. What follows is an exploration of these two transformations. Neither represented a 'natural' shift, but instead were shaped violently, moulded by the demands and fears of East India Company administrators, army officers, medical surgeons and evangelical missionaries.

This shift was closely linked to officers' and administrators' perception of their own European rank and file. This understanding placed the men at the very bottom of an imagined, moral hierarchy. Indeed, as late as the twentieth century, the recruits of the eighteenth century were looked back upon as the 'scum' of the nation, a rowdy assortment of reprobates, drunkards and pickpockets thrown together with the labouring poor who completed their ranks.[3] In fact, labourers composed the majority of troops and, as Stanley has pointed out, there was a certain degree of respectability associated with service in the Company's army.[4] However, in the vision of the officers and administrators who made vast pronouncements on the men, this fact was largely disregarded. It was a fear of potentially upsetting the men – provoking not only their ire, but raising the spectre of possible mutiny, which, combined with the very real and pressing fiscal dilemmas which the East India Company faced in the late eighteenth century that informed military and medical responses to venereal disease control.

The army's attitude towards its own soldiers changed very little over the course of the nineteenth century. Among Company officials, the prevailing view of the soldiers as low-class and 'degraded' brutes persisted. Officers saw the men as volatile and violent and, as such, believed them to require careful handling. The men's supposed sexual 'needs' were one facet of their presumed character that merited careful attention. Army officials assumed that in the absence of a wife, an alternative would need to be found to satisfy the soldier's sexual wants and urges. Without such a 'release', it was feared that the men would otherwise resort to 'dangerous' or 'deviant' sexual behaviour – masturbation or homosexuality, in addition to rape or assault. The women living in and around the camps who served as prostitutes would form a large part of this 'solution'. Yet rising levels of venereal disease among the men soon complicated this situation.

This chapter begins with a brief exploration of the development of the Company armies in India. It pays particular attention to the growth of the European contingent within these forces as the Company's political and territorial power grew in the eighteenth century. While an unflattering view of the common soldier was certainly not unique to Company officials, this view fit particularly well with the overall ethos of the Company, who, led by profit, sought to maintain power and trade in India as cheaply as possible. As such, as Company rule spread and further expense was required, it was uninterested in any investment in the European soldiers except that which was absolutely necessary. Dealing with the men as 'degraded' and morally irredeemable and instead seeking to marshal their brute strength was the shrewd, more economical option that both Company and Crown chose to pursue. The low pay and benefits the soldier received as well as the restrictions imposed on those who surrounded him were perfectly in keeping with this unwritten policy.

An exploration of the interlinked issue of the relationships that existed between European soldiers and officers and Indian women in the late eighteenth and early nineteenth centuries follows. These relationships, while initially encouraged by the Company, would come to be seen as a burden and were accordingly discouraged. This transition was not simply the result of changing racial attitudes in metropolitan Britain; instead, it materialised as a result of very specific factors in India and London. After 1800, wary of having to provide for the ever-expanding families of its soldiers, the Company began to discourage the provision of allowances and pensions for the wives and children of soldiers. Racial difference provided one convenient excuse for paying these companions less money, or indeed, none at all. There were some officials who argued that supporting Indian wives and mistresses made sound economic sense. They asserted that Indian wives and mistresses should be encouraged as they provided the much-needed 'comforts of home', such as improved diet and greater attention to dress, and engendered soldiers who were healthier and better disciplined than their unattached peers. Accordingly, for much of the nineteenth century, this transformation of relationships was uneven. However, by the century's end it was almost entirely complete and Indian companions were relinquished in favour of 'regulated prostitutes'.

Desperate to ensure that venereal disease did not fatally weaken the army's defensive capabilities, regimental surgeons sought inexpensive methods of treatment and prevention for syphilis and gonorrhoea which would have the least impact on the men. The system of regulated prostitution was the product of these fears. Supporters of this system argued that it would reduce venereal disease by providing the men with a 'healthy' pool of women who were closely monitored for any signs of disease. By this system, surgeons hoped to minimise the number of men rendered unfit by instead shifting the burden and risks associated with treatment to this group of women. What is more, it would avoid 'provoking' the men. However, from the start, the system failed to achieve its goals. Seeking explanations for this failure, surgeons and commanding officers pointed not only to the women under its remit, but to other women who did not fall under its net. These included dancing girls and courtesans. These groupings extended far beyond the very base 'prostitute' classification and did not easily conform to prevailing British models of sex and morality. Although it is unlikely that many women in these categories would have considered themselves to be of 'ill repute', British observers, having little to compare their roles to within British society, began to fuse them all under the misleading title of 'prostitute' in the 1810s and 1820s. In many ways, this transformation reflects the contemporary context of Orientalist approaches and interpretations of Indian society and traditions. As was the case with British understandings of Indian laws – Hindu or Muslim – many officials and observers adopted a flattening approach to certain Indian roles and traditions.

Moreover, with limited access to certain practices, such as temple dancing, European understandings were passed down based on only the briefest glimpses of the practices under question. The medical, missionary and military targeting of such groups, combined with political and economic changes, meant that by the mid-century, these women came to be designated simply as 'prostitutes' by the colonial state.

This chapter suggests that the moral and medical boundaries erected in the early nineteenth century around women deemed 'prostitutes' paved the way for their later condemnation and criminalisation as well as dramatically re-defining their social and legal status. When the Indian Penal Code of 1860 was enacted, the Indian 'prostitute' was assumed to be a member of the 'criminal' category and targeted as such by a number of sections in the Code.[5] The women's separation from what was deemed 'respectable' Indian society and insertion in a 'criminal' category was crucial to surgeons and commanding officers' attempts to justify the invasive and socially disruptive methods they proposed to control venereal disease among the European soldiery. Officials presented women deemed to be 'prostitutes' as a 'threat' to the military (and therefore to the Company's overall security). To achieve their goals, military and medical officers often adopted the moralising tones of missionaries (who stressed the women's supposedly innate immorality).

The transition from the earlier social and sexual mores, which had encouraged long-term relationships between Indian women and European men, to a more puritan emphasis on distance and control has been insufficiently interrogated and is often regarded as simply part of a broader shift towards racialist or racist constructions of colonial rule. This chapter explores some of the complexities of these new constructions, arguing that the changed view of Indian women and indeed society, which emerged in the mid-nineteenth century, was not one wholly imported from Europe. Quite the contrary, it developed in large part out of the daily interaction and tensions between European medical men and military officers on the one hand and European soldiers and Indian women on the other, surrounding a range of issues from remunerative support for wives and widows to the treatment of venereal disease.

The East India Company and its army

Formed by a group of wealthy businessmen intent on breaking the Dutch monopoly on the spice trade, the East India Company received a Royal Charter in 1600 from Queen Elizabeth I, which effectively gave it a monopoly on all British trade in the East Indies. The Company quickly began to establish trading posts (known as 'factories') on the coasts of India. In an effort to secure both trading and security privileges, the Company moved to form strategic alliances with Mughal rulers and a number of 'successor' powers.[6] With the Company's expansion in the eighteenth century, the three largest of these factories were transformed into walled forts. In turn, these forts developed into the three Presidencies from which India was administered. An assigned Company president had effective authority over Fort William in Bengal, Fort St George in Madras and Bombay Castle. Each of the presidents, or governors, oversaw the areas with the assistance of a Council, composed of senior Company servants. These Presidents ruled with relative autonomy – the aim being to oversee and protect the Company's commercial interests in the area surrounding its factories. This degree of independence amongst the Governors of Bombay and Madras persisted even after the Regulating Act of 1773 elevated the status of Bengal over Bombay and Madras.[7] By the late seventeenth century, surrounded by competitors and, at times, local rulers hostile to their presence, the Company began to develop more actively small armies to protect their interests.

The three Presidency armies had humble origins, growing from an 'ensign and thirty men' in Bengal, a detachment sent to garrison Bombay and, finally, the re-categorisation of factory door-keepers and soldiers into a company at Madras.[8] Simply put, the earlier groups of armed guards who protected the fort and factories grew and in time transformed into the presidency armies. These mercenary forces were largely drawn from the local population in the areas surrounding the main factories. After 1754, smaller units of Crown troops joined the Company forces. The threat of hostilities with France (both in Europe and on the subcontinent) and the opportunities presented by declining Mughal power proved a potent combination that prompted the expansion of Company operations. As the Company grew and increasingly intervened in the political and military balance of power in India, the size and importance of its armies grew apace. Like the differences that persisted in the governance of the three presidencies, each of the three armies maintained distinct policies on everything from recruitment and composition to promotion and benefits. In 1796 (following simmering officer discontent that erupted into a minor mutiny over the issue of *batta,* or field pay), the Company introduced a series of reforms in an attempt to rationalise these variations.[9] Further alterations were attempted in 1824; however, anomalies remained until the Company's Presidency armies were absorbed into the British army after 1858.[10] In a similar manner, stereotypes (and nicknames) persisted about the kind of men serving in each of the armies. The Bengal 'Qui Hai' was arrogant, while the Madras 'Mull' was lazy and the Bombay 'Duck' had fewer benefits and comforts than the other two.[11] Petty rivalries persisted between the three armies for much of the century.

Its 'sepoys' – Indian troops drilled in a European style and led by European officers – comprised the bulk of the Company's army in India.[12] The French Governor-General Joseph Francois Dupleix was the first to raise sepoy battalions in southern India in 1748. The East India Company army under Major-General Stringer Lawrence followed suit a few years after.[13] The Company recruited these men from regions in northern India that had historically served as military catchment areas, namely Bihar, Benares and Oudh.[14] The Company also maintained a smaller contingent of European soldiers; however, for most of the period until the mid-nineteenth century, sepoys outnumbered European troops by a ratio of eight to one.[15] This was largely due to the relative expense and higher susceptibility to disease of European soldiers. A 1781 letter from General Stubbert to Warren Hastings assured him that even in the most healthy seasons, an eighth of the European force was rendered unfit for service due to illness.[16] In 1765, the number of sepoys employed by the British was roughly 9000; by 1808, this number had grown to over 155,000.[17] By way of comparison, in 1790 the number of British forces serving in India was roughly 18,000.[18]

While this imbalance persisted, with the spread of its power, the Company recruited increasing numbers of European soldiers from the United Kingdom to serve in India. As noted above, these soldiers were primarily drawn from the British or Irish working classes – predominantly labourers. While much has been made of the position of Irish soldiers in the army, by the mid-nineteenth century, their dominant (at least in relation to the overall British population) position within the army had evaporated due to famine and emigration.[19] Recruits were mostly the young, working poor, with an average age of between 15 and 19 years and a daily wage of about one shilling.[20] With this wage, it seems unlikely that many men, save those from the lower classes, would be enticed to register for army service. Many of the men were illiterate and an even greater number were in poor physical condition even before their arrival in India.[21]

The recruiting process itself was heavily reliant on the use of alcohol. While no recruit could be officially attested until 24 hours had passed since his enlistment (ostensibly to prevent rash, ill-chosen and drunken choices), recruiting sergeants often skirted this requirement by keeping the men in a constant state of intoxication until the 24 hour had passed.[22]

A small percentage of the men were married, yet very few could afford to pay the passage for their wives to join them in India. Moreover, only a limited number of places for the wives of soldiers were permitted for each regiment. For families to *qualify* to enter what was a draw for one of these spots (12 places for each regiment travelling to India), the couple had to have sought official Company approval for the marriage. Those wives whose names did not come up in the lottery were ordered to return to their home parishes.[23] Some, like the four young married women who Sergeant John Pearman remembers smuggled aboard the *Thetis,* the ship that carried him to India from Gravesend, did so without obtaining Company leave. As such, these women not only risked deportation, but as the Company did not recognise their status as wives, they received no official provision.[24] Even when the number of women permitted to join their husbands was increased in the second half of the nineteenth century, the proportion that actually did so remained low. In 1860 only 6.5% of soldiers elected to bring their wives to India with them.[25]

Despite the fact that 'native' sepoys made up most of the army in India, it was the health of the European soldiers that the Company was most concerned with. While a whole set of assumptions regarding the 'character' of the sepoys was later developed (which among other things stressed certain groups' 'martial' or war-like characteristics),[26] the beliefs held regarding European soldiers were often far less flattering. As David Cannadine has reminded us, it is too simplistic to view the empire just as a complex racial hierarchy – class and social structures played a critical role in its formation and operation.[27] Army and government officials (themselves drawn primarily from the upper and upper-middle classes) held firm stereotypes about the 'degraded' class of men from which the European soldier was drawn. Nor was the army's administration unduly concerned about this lack of 'refinement' amongst its ordinary soldiers. Not only were both Crown and Company unwilling to offer greater pay to attract a higher calibre of soldiers, but for most of the nineteenth century, they also resisted all pressure to invest in the education of the men. Too much refinement was explicitly deemed to be *undesirable* among ordinary soldiers. The Duke of Wellington, Arthur Wellesley, was said to have been distinctly uninterested in promoting education for the working classes.[28] As later chapters show, while there were some moves to introduce more 'wholesome' entertainments, such as reading or coffee rooms, in to some cantonments, these were limited. In the early nineteenth century the introduction of such novelties depended very much on the whim of the particular commanding officer.

Company officials constantly debated the composition and wants of the troops. One historian has suggested that as Company forces were recruited and offered higher pay, they were of a greater moral calibre than their Crown counterparts and often sought to establish a life for themselves in India.[29] There is some indication that they were more likely than Crown soldiers to establish 'roots' in India either by marrying or co-habiting with native women.[30] Significantly higher pay might perhaps have allowed Company men more privacy to conduct their personal lives; however, it appears that neither force was paid enough to absolutely ensure this small comfort. Most soldiers, without the financial means to do otherwise, lived communally without privacy in the barracks, surrounded by both their women and children.[31] Despite the presence of established army families (albeit mostly 'unofficial' either in the form of unrecognised British or Indian wives and mistresses), from the late eighteenth century, the argument was made that due to the 'class' composition of British soldiery in India, they had an innate inability to control their sexual impulses. There appears to have been little evidence to support this assertion. In all likelihood, the truth lay closer to the fact that administrators were unwilling to admit that in order to reasonably afford the maintenance of a family, the men's salaries would have to be substantially increased. The eighteenth and early nineteenth century saw Company officials performing a delicate balancing act; on the one hand, they were keenly aware of the benefits which accompanied 'native' wives and mistresses (ranging from the men's improved health and discipline to greater 'knowledge' of the country), but on the other, they were reluctant to shoulder the costs of these relationships. In resolving to avoid the latter, commanding officers, surgeons and Company officials definitively shaped sexual and social relations in British India.

This conception of the 'degraded' soldier would prove resilient and long-lasting. Writing about the British army in 1868, Sir Charles Trevelyan insisted that military efficiency depended on the intellectual capabilities of the higher ranks and the brute strength and agility of the lower ranks.[32] Even when officers later expressed a desire for a better *class* of recruits, they were not searching for those of a higher social class but a better selection from the trades already preferred by the army.[33] Yet certain elements of their 'natures' – their supposed degeneracy and immodest attitude towards drink and sex – were seen as hazardous, endangering, as they did, their very health. This is perhaps especially ironic given the heavy reliance on alcohol during the recruitment process.

As the Company grew, it made more extensive physical and sexual provisions for its army based on this narrow understanding of the men's character. By the mid-eighteenth century, many cantonments had expanded to include regimental and central bazaars so that a soldier would not need to venture into the native bazaars to satisfy his 'needs'. By the late 1770s, some cantonments also included a *lal* bazaar, which essentially amounted to a regimental brothel.[34]

Begums and Bibis

Despite the lack of material support from the Company, relationships between European men and Indian women continued to be fairly common in eighteenth- and nineteenth-century India. Until the mid-nineteenth century, the East India Company was extremely wary of allowing any increase to the number of European women in India, fearful that any such change might pose a threat to the stability of its relationships with Indian rulers. The Company therefore required all European travellers to obtain explicit permission prior to their departure for India. In January 1755, the Company ordered Miss Campbell, one such woman who failed to obtain this dispensation, to return to England at the expense of the owners of the boat which had brought her from Madeira to Bengal.[35] As a consequence of this approach, the number of European women in India was extremely limited until the late 1830s. During this period the prevailing, if unofficial, social standard for European men was to enter into relationships of concubinage with Indian women. In the seventeenth century, the Company's Court of Directors went so far as to encourage its soldiers in Madras to marry Indian non-Christian women in an attempt to prevent them from taking Portuguese wives, whose Roman Catholic religion was considered potentially more dangerous to the Company's stability than either Hinduism or Islam.[36] Until later theories of race and morality came to dominate policy, these relationships existed in nearly every area of India where East India Company men lived.

Although no documentation survives which expressly confirms official sanction for these relationships in the early nineteenth century, we can infer tacit acceptance from certain discussions on the health of British soldiers. One surgeon proposed that to reduce disease and promote efficiency, officers encourage their soldiers to embark upon relationships with native women, as he insisted they were a proven sobering influence.[37] The same officer went further, expressing his conviction that European women did not possess strong enough constitutions to withstand the various climatic challenges India posed to their health. He advocated that married men, should their regiment be posted to India, be attached to a different corps and be permitted to stay with their wives in Britain. On the other hand, he argued that single men should be sent to India and, on arrival, be partnered (for their health and sobriety) with Indian women.[38]

The Madras Medical Board concurred with these assertions. In its 1810 proposal to the Governor it noted that those men who had attached themselves to native women were usually kept free from venereal disease, and had more attention paid to the 'comforts conducive to health' than was the case of men with European wives.[39] These medical men saw European women as more of a hindrance than a help, as they thought their 'constitutions' not strong enough to endure India; accordingly, the majority of such women were 'lost to the country'.[40] In addition, although not explicitly mentioned in this proposal, European women were more expensive to maintain – not only was it felt that they would require certain comforts and amenities, but the costs of 'appropriate' housing, food and medical treatment, all needed to be taken into account. Despite enthusiastic backing from the Board, the Company never officially approved the Madras plan and supporters of such proposals were forced to look for alternatives.

Marriage between a European man and an Indian woman could only take place if the woman was (at least nominally) a Christian. Therefore, the bride's conversion and baptism often preceded a church ceremony. In 1755, the church register for Calcutta shows that of the thirteen recorded marriages, seven list the bride simply as 'countrywoman', the common parlance used to indicate her non-European origin.[41] While official church marriages were perhaps less pervasive in comparison with the number of unofficial relationships, both were socially accepted. In a 1781 letter to his brother, Samuel Hickson, an officer in the Company's army in Madras, gave a frank insight into the prevalence of such relationships.[42] Hickson revealed that most of his compatriots, after arriving in India (with every intention of eventually returning to England), had subsequently embarked on relationships with Indian women and chosen to settle in India with their wives (or mistresses) and children. Hickson noted that 'very few of those, who ever enter into any of these engagements ever think of going home afterwards'.[43] In his battalion, there was only one European, besides himself, who was single (and this man was apparently doing everything in his power to rectify that situation).[44] Hickson described in detail his Drill Sergeant's courtship and marriage to an Indian girl who he estimated to be about thirteen years old. His account of the ceremony highlighted the intermixture of cultures, with a traditionally dressed and bejewelled Indian bride, an English church ceremony conducted within Fort St George, followed by music and dancing.[45] Indeed, this practice continued through the century, though on a decreased scale. The diary of one Sergeant recalls a fellow soldier in 1849 who, when the regiment was due to return to Britain, opted to remain in the country with his 'black wife'; however, by mid-century the numbers of such wives had greatly reduced from that seen in Hickson's time.[46]

Soldiers were especially encouraged to seek their brides from the orphan schools – those established in the late eighteenth century for the 'orphaned' children of European soldiers. Often, the term 'orphan' was a euphemism to describe the illegitimate children of European fathers by Indian or Eurasian mothers.[47] The schools developed out of the Military Orphan Society, which was established in 1782 to support the Eurasian children of European officers. In time, the children of 'poor whites' who could not afford to send their children to Europe to be educated became the main focus of these schools. By 1803, one such school, the Calcutta Lower Orphan School had over 500 boys and girls on its register.[48]

The schools housed and fed the children, raising them as Christians and teaching them English and the skills thought most 'useful' to their upbringing; boys learned vocational trades and girls domestic skills. These were not simply benevolent institutions; their aim was to shape the children into useful members of society. This process was seen to begin with their removal from the 'dangerous' and 'corrupting' influences of barrack and bazaar society.[49] The fashioning of young girls into marriageable, marketable wives was further encouraged by the regular balls hosted by the schools which served as an occasion to introduce the soldiers to all girls of 'appropriate' age.[50] There was no restriction on the men marrying girls from the schools. All that a soldier needed to embark on such an engagement was his assurance that he was single, along with a letter of permission from his Commanding Officer.[51] The returns of one such school in Bengal show that between 1800 and 1818, an average of just over 20 girls married European men each year (see Table 1.1).

Permission to marry women from the orphan school was usually readily granted as it was, in the words of one observer, seen to '... steady the soldier and induce him to habits of sobriety'.[52] The same observer

Table 1.1 Returns of the number of females who have been married out of the Lower Orphans [sic] School, 1 January 1800 to 31 December 1818

To non-commissioned officers & privates of His Majesty's services		To non-commissioned officers & privates of the Hon'ble Co's services		To persons not in the service of His Majesty or that of the Hon'ble company	
In the year	**Number**	**In the year**	**Number**	**In the year**	**Number**
1800	"	1800	14	1800	1
1801	"	1801	13	1801	1
1802	"	1802	12	1802	1
1803	"	1803	9	1803	1
1804	"	1804	2	1804	2
1805	"	1805	10	1805	4
1806	"	1806	10	1806	1
1807	"	1807	16	1807	"
1808	1	1808	20	1808	2
1809	4	1809	27	1809	2
1810	22	1810	16	1810	5
1811	4	1811	10	1811	4
1812	1	1812	16	1812	3
1813	1	1813	19	1813	2
1814	5	1814	7	1814	2
1815	18	1815	19	1815	2
1816	2	1816	26	1816	1
1817	6	1817	17	1817	6
1818	1	1818	11	1818	1
Total	**65**	**Total**	**274**	**Total**	**41**
Average number per year during 19 years	3 ½	Average number per year during 19 years	14 ½	Average number per year during 19 years	2 ¼

Source: *Returns of Marriages from the Lower Orphan School from January 1800 to December 1818*, Bengal (Separate) Military Collections, 1818. APAC L/MIL/5/376, col 3.

went on to note that he had it on good authority (from a number of regimental officers) that many of the men, after marrying girls from the orphan school, had become '... the steadiest and most orderly in their regiments'.[53] In line with this respectability, events such as monthly balls continued for some time in many stations. In his diary entry for 1846–1848, Sergeant Pearman of the 3rd Light Dragoons recollects that there was a monthly 'Bon Ton', or dance to which '... all the females of the station [were] invited ... half caste and all, so long as they were wife of a soldier'.[54]

While comments such as these (and the records of the orphan schools) help to form part of the picture of the contours of relationships between soldiers and Indians and Eurasian women, different sources are available for examining the lives of men of the officer class as well as those civil administrators. The wills of officers in Bengal in the late eighteenth and early nineteenth century provide further evidence as to the widespread nature of these relationships.[55] It is not terribly surprising that this source is not available for soldiers, as few of the men accumulated enough material resources to bequeath to the family or friends who survived them. Officers' wills, and those of higher rank, which specifically name Indian women indicate that most of the men involved in these relationships were monogamous (or serially monogamous), although some had multiple wives or mistresses. One such man was Matthew Leslie, formerly of the Civil Service, who served as collector and judge (in Patna and Benares). Leslie's 1804 will outlines the substantial payments made on his death to his three wives, one mistress and six children.[56] The will highlights what appears to be a structured ranking of Leslie's female companions. The most highly ranked, Zehounun Khanum, is identified by Leslie as the former wife of Meer Mahomed Hussein Khan. In his stress on identifying her former husband by name, Leslie draws attention to her 'respectable' status. He awarded Zehounun the most money of the four women – 20,000 sicca rupees. To 'Keera Beeby [sic]' and 'Zebon' (for whom no surnames were given), Leslie left 12,000 sicca rupees each. In addition to the cash, he gave each of the women a house in Patna.[57] To 'Assoorum', the woman who appears to have been an unofficial wife (or mistress), he left 3,000 sicca rupees. Leslie does not refer to the women specifically as his 'wives' in his will. Instead, he uses the amorphous term 'my girl' for each. Indrani Chatterjee has interpreted such phrases as euphemisms that instead indicate a woman's status as a domestic slave.[58] Although there is no mention of a fourth wife (presumably Assoorum), family records would appear to indicate that Leslie *married* three of the women.[59] Furthermore, while it is possible that Keera Beeby, Zebon and Assoorum entered Leslie's household as slaves or staff, it is highly unlikely that the same was true of Zehonun Khanum.

In other cases where men were not officially married to their Indian companions, they refer affectionately, but obliquely, to them in couched terms. Thomas Naylor, a major in the infantry, left 4,000 rupees to his pregnant 'female friend', Mukmul Khanum, as well as the bungalow they shared in Berhampore[60] (fully staffed by their male and female slaves) and a 3,000-pound provision for their unborn child.[61]

In 1810, Captain Thomas Williamson, a former officer in the Bengal army, published his *East India Vade Mecum,* a travelogue considered essential reading for all gentlemen heading to India.[62] The *Vade Mecum* contained much meticulous detail regarding the maintenance of Indian wives and mistresses. Williamson outlined the costs for supporting an Indian mistress (40 rupees per month, or roughly £60 per year) which, he insisted, was 'no great price for a bosom friend; when compared to the sums laid upon some British *damsels,* who are not always more scrupulous than those I have described'.[63] Williamson estimated the number of available said British 'damsels' in Bengal to be no more than 250. He asserted that nine out of 10 women in such relationships were Muslim and insisted that the small percentage who were not were Portuguese Catholics. The majority of these women never converted to Christianity, nor did they give up their own traditions.[64] Williamson stated his preference for the Muslim women, asserting that the Portuguese '... have no scruples as to what they are to eat and drink; many of them, indeed, can manage a bottle as well as any man in the kingdom'.[65] Indeed, as Chapter 4 shows, the 'unfeminine' drinking habits of European women were seen to directly threaten the delicate balance of life in the cantonment. Throughout his description, however, Williamson was careful to point out that both Indians and Europeans viewed such concubinal relationships as 'equally sacred' to marriage.[66]

Similarly, published personal recollections such as Captain Bellew's 1843 *Memoirs of a Griffin* fondly recount such liaisons. Bellew wrote of his General's attachment to Sung Sittara Begum (the 'Queen of Stars'), which he describes as a 'harmonious' and 'enduring' union which ended only with the begum's death.[67] In her 1850 memoir-cum-travelogue *Wanderings of a Pilgrim, in Search of the Picturesque,* Fanny Parkes recalls meeting Colonel Gardner's Begum Mulka and notes (with a touch of jealousy) the enduring dedication the couple felt for each other.[68] While many of these wills and memoirs implicitly acknowledge that the couples were not officially married, there is no indication that any of these men felt that there was anything immoral or untoward about their relationships.

Even in the midst of such relatively permissive attitudes, Indian wives and mistresses and their mixed-race children encountered a measure of resistance from Anglo-European high society. Williamson notes that a Eurasian woman (even if she were the daughter of a man of the highest rank who was married to a man of similar rank) would not be invited to parties hosted by the Governor.[69] However, throughout the nineteenth century, it was not the sexual relationships between European men and

Indian women which were in themselves viewed as problematic; it was their mixed-race progeny and the expectation that they would demand assistance from government which provided much of the fuel for social and economic tensions. The children posed a particular threat due to the belief that they might attempt to claim financial or political rights based on their father's position.[70] The Company, wary of having to potentially support thousands of native widows and their mixed-race children, began to discourage any expectations that either might have for relief.

Cornwallis' reforms of 1791 excluded persons of mixed race from holding either political or military office with the Company (with the exception of those serving as drummers, fifers or other musicians). This was developed on the back of similar policy enacted within the Royal Army a few years earlier. Hawes suggests the rationale behind this rested with a fear of political reliability, rather than race or legitimacy.[71] Whatever the reasoning, however, the reforms ensured that henceforth the respectable military and civil positions that the sons of mixed relationships had previously filled were closed to them. This fact notwithstanding, in 1798, of the 44 wills made by Europeans in Bengal, 13 still made provisions for their Indian mistresses and the children born to them.[72] Despite the knowledge that their children would not be entitled to hold an office within the Company, both they and their mothers continued to be formally acknowledged as the men's legitimate heirs and beneficiaries. Nevertheless, Cornwallis' reforms forced a subtle, but significant shift in the perception of these relationships which would eventually contribute to their decline.

The growth of the Eurasian population raised the additional question of what was to happen to families when soldiers (and officers) returned to Britain. In 1817, Government asked the Advocate General for his opinion as to whether there were any legal conditions that would bar the Indian and Eurasian companions of soldiers and their children from being sent to England. He replied that, provided the men were legally married, he knew of no restriction which could stop them from sending their wives and children to Europe.[73] Furthermore (and perhaps of greater concern to the Company), he noted that he could find no impediment which would stop the men sending their children, even those born *out of wedlock* to Europe, if they so desired. While the Advocate General was rather vague about who was responsible for the all-important matter of *paying* for such transport, the implications for the Company were potentially enormous. Indeed, these concerns did not disappear. One later proposal from the Adjutant General suggested a different alternative for those men who were married to Eurasian women. These men, the letter suggested, should be given the option to 'volunteer' to continue their service in India when the rest of their regiment returned to England as it would be '... unkind to turn such women [i.e. their wives] upon the world in England, where they have no home, no friends, not even a work house they can claim a refuge.'[74]

In these debates we can see that the Company was increasingly unwilling to bear the expense of maintaining such families, both official and unofficial. As outlined in the introduction to this book, the unstable state of the Company's finances meant that economy was sought wherever possible. 'Superfluous' expenses – like support for the families of soldiers – threatened to push the Company further into debt. A fierce dispute erupted after the return of one regiment to England left the men's female companions stranded in Bengal. In 1817, the army ordered His Majesty's 66th Regiment to embark for St Helena from Bengal. The regiment had earlier served in Ceylon, where many of the men had begun relationships with local women. When the regiment relocated from Ceylon to Bengal, the commanding officer had 'permitted' these women and children to accompany them. However, when the men prepared to leave Fort William for St Helena, the 55 (unmarried) women and 51 children attached to the regiment faced the prospect of destitution.[75] The suggestion that they be allowed to travel to England with their partners never arose. Instead, the debate centred around the best way to return the women to their families in Ceylon and where the responsibility for covering this expense lay. Colonel Henry Torrens, writing from Army headquarters at Horse Guards in London, sought to underscore his insistence that it was *not* the British government's responsibility to pay for these 'unofficial' families' return to Ceylon and insisted that such costs be borne by the Company.[76] The British government would only provide for those children born *in wedlock* of *European* mothers. Torrens went on to stress that it was 'reprehensible' that the commanding officer had ever permitted the women to accompany the corps (although he did not challenge the officer for allowing the men to embark on the relationships in the first place). After a protracted debate, the government in Bengal decided that the women and children would be returned to Ceylon and the cost paid for out of the Company's budget for 'extraordinary' expenses.[77] Debates such as these hardened the Company's resolve to avoid any unnecessary expenditure when it came to the families of soldiers.

As troop numbers increased and military budgets stretched, the potential cost of supporting the men's families must have been ever-present in the minds of budget-conscious administrators. In these debates, we see that the very idea of 'family' was malleable and largely dependent on the particular needs of the army. On the one hand, the Company, in carefully restricting the number of recognised 'official' European wives, encouraged a (faux) bachelor army in India. As the men persisted in establishing long-term relationships with Indian wives and mistresses, they found the Company could act equally arbitrarily – forcing their separation when regiments departed India.

Concerns over costs in cases of 'abandonment' were relatively minor in comparison with the potential expense of extending the provision of monthly allowances (paid to men with families) to Indian and Eurasian companions. In 1797, the military allowed for a small subsidy to be paid to the limited number of wives of European soldiers who had accompanied their husbands from Europe. Even in the case of these women, if their circumstances changed – and they were, for example, widowed, they needed to swiftly find a new husband as they would go 'off pay' in as few as six months from the death of their soldier husbands.[78] Sergeant Pearman recalled returning to quarters after the Battle of Sobraon[79] in the First Anglo-Sikh war, to note that the war had made 14–15 new widows in his regiment. Most of these women remarried within a month of the Regiment's return. Pearman despairingly noted that some had three or four previous husbands, though the (new) Mrs Sergeant Gooderson, who now entered into her sixth marriage, held the record.[80]

Even this small support was withheld from Indian wives (or mistresses, who of course represented a much more significant number). It was only in 1824 that the military granted these wives any allowances at all. At this time, officials suggested a monthly stipend of 5 rupees for European wives and 4 rupees for 'half cast' wives, with the specific proviso that 'women of colour' were exempt.[81] Even this allowance raised protests from cost-conscious administrators. Madras Governor Thomas Munro objected to the proposed changes, arguing against any remittance for Eurasian wives and children.[82] First, he stated, the expense incurred through such a policy 'will be great beyond what we can foresee' and beside this he noted, '... we shall be embarrassed by the accelerated increase of a race for whom we cannot provide and whom we shall have taught by indulgence not to provide for themselves'.[83] This was, of course, somewhat ironic coming from Munro, who himself had a Eurasian family. However, it seems that in this instance, his reputation as a keen balancer of budgets won out. Moreover, Munro warned that even if only one-third to one-half of their European soldiers married Eurasian women, it would mean 10,000 to 15,000 wives (not to mention children), to support at public expense. Instead of encouraging this 'disastrous' situation, Munro suggested that no aid be bestowed to these wives and families beyond what was already provided by charitable institutions. Munro also believed that the cost of paying for *European* wives and children of soldiers had become too great a burden on the Company. Therefore, he proposed, their numbers should be severely restricted. He was first and foremost concerned with cost, then with military efficiency.

Throughout the early nineteenth century the Board of Directors frequently returned to the debate over the expense of soldiers' families. In 1770, the Company established Lord Clive's Pension Fund to provide for soldiers and, in the event of the men's death, their widows. As with the Military Orphan Societies, the Fund performed many of the functions of a state (such as education and financial support) before the colonial state itself was formally established.[84] To apply for monies from the Fund, a widow was required to apply first to the Military Department in the presidency where she was resident. The Department would discuss her request before sending its recommendation to London where the Directors made a final decision. While it appears that the Directors were initially generous in granting these requests, after 1800, as the Company attempted to cut costs, they increasingly denied relief to women who were not of 'pure' European descent.[85] The Company side-stepped the issue of supporting the Eurasian and Indian wives of soldiers and officers in 1808 by noting that the Fund had been intended for those officers, soldiers and their widows *resident in* Great Britain or Ireland, and therefore, wives living in India had no right to make a claim.[86] Moreover, the author continued, as the demands on the Fund were much greater than its income was able to provide for, the Company had been forced to pay a large amount *out of its own assets* to provide such pensions.[87] As a commercial operation whose prime goal was profit, such an 'imposition' must have been abhorrent to Company Directors. Accordingly, lawfully married women of Indian or mixed parentage were found ineligible for the Fund for some time to come.

As noted above, a change to this policy came in 1824 when Eurasian widows were granted similar rights to European women for relief. However, these women received greatly reduced allowances in comparison with their European counterparts. The solution that the Company appeared to adopt was to discourage all claimants by requiring them to produce physical, administrative proof of the legitimacy of their relationships. Women were required to submit their petition along with written confirmation from a Church of England Chaplain that they were legally married (a statement that was, of course, impossible to produce for the many unofficial wives).[88] An 1829 letter reveals that even after these limited changes were made, the Madras Fund continued to reject legitimate petitions submitted by Eurasian and Indian widows.[89] It was not simply the issue of pay which Indian and Eurasian women had to fight for. A Bengal military letter revealed that the government there also begrudged the costs attendant on women's medical treatment when sick; it was only in 1826 that it announced that the 'indulgence' of being received into hospital when sick would be extended to the Eurasian wives of soldiers.[90]

Nevertheless, this administrative burden did not prevent most Indian and Eurasian wives from applying for assistance. The sheer volume of petitions filed by these women serves as further proof of just how common such relationships were.[91] It also suggests that the women had a strong sense of what their rights were by virtue of their relationships with European men.[92] The ongoing demands made on Company resources for support continued to perplex officials attempting to balance budgets. In a fresh attempt to discourage claimants, official bureaucratisation was taken a step further in the 1840s. At this time, additional documentation was required (in addition to the marriage certificate), which spelled out the woman's exact descent or educational status (see Table 1.2).

Any inaccurate declaration made on these certificates was grounds for the automatic dismissal of a woman's claim. Where the government could avoid paying for allowances and pensions to the wives and widows of soldiers and officers, they did so. In May 1826, Corporal Francis McAlister, a man deemed to be of 'exemplary good conduct', married the daughter of a Eurasian mother by a Portuguese father. Shortly after their marriage, McAlister applied on his wife's behalf for her monthly allowance of 4 rupees (which he believed her to be entitled to as the child of a 'native woman' by 'European father'). McAlister's request was denied. The Military Fund argued that she did not come 'within the pale of the regulation' as her father was 'native Portuguese' and therefore not 'European'.[93] The categories that encompassed 'race' and what it meant to be 'European' were negotiated, shaped and hardened by officials seeking to limit their exposure to the potentially vast number of claims that could have been made for financial assistance. By setting such a draconian example in this refusal, the Company no doubt hoped to deter future claimants.

While the strongest criticisms of these relationships came from those managing costs within government, after 1813 evangelical missionaries echoed these critiques.[94] These missionaries sought to portray the 'morally corrupting' influences of India in their description of European relationships. Just as the Company cited racial differences when these could be used to justify not providing the same material support to

Table 1.2 Certificates of descent and education establishment

No 1 European

Certified that A. B. wife of Gunner C.D. of the __________
Company __________ Battalion artillery was educated at the (here enter Regimental or other school in which she was educated, and the period she was there) and that her descent is as follows.

Her father was a European, his name was F.E. a sergeant in the Regiment of Her Majesty's service.

Her mother was an Indo Briton and her Maternal Grandfather a European, his name was G.H. a Gunner in the Battalion of Artillery.

C.D.

Commanding

No 2 Indo Briton

The same with descent as under Her Father was a European his name was E.F. a private in the Regiment of Her Majesty's Service.

No 3 Indo Briton

The same with descent as under Her Father was an East Indian whose name was E.F. a cabinet maker at Madras, and her mother was also an East Indian.

Both her paternal and Maternal Grandfathers were Europeans, the former named G.H. was a private in the Grenadier Company of Her Majesty's Regiment and the latter named I.J. was a Gunner in the Company Battalion artillery.

No 4 For those who have not been educated at Regimental or other established military schools

Certified that A.B. wife of Gunner C.D. of the Company Battalion Artillery has been examined by me, and that I consider her to have received such instruction, as the daughter of a Christian soldier should receive.

C.D.

Chaplain

Certified that the descent of A.B. wife of Gunner CD of the Company Battalion artillery is as follows.

Here enter the descent as above, as the case may be.

Note Both certificates in No 4 to be written on the same sheet.

Source: *General Order by Major General Sir Robert Dick, Commanding the Centre Division of the Army, 24 June 1841*. TNSA, Madras Military Consultations, February 1841.

Indian and Eurasian companions as was provided for European women, missionaries stressed the supposed immorality ever-present across India in order to shock (and perhaps, more cynically, elicit financial contributions from) their congregations back in England. It is important to remember that most missionaries sent their 'observations' and letters back to London so that these could be read or distributed to congregants at their sponsoring mission. In 1814, the Bombay Bible Society launched a public assault on the manner in which the children of European fathers (by both European and Indian mothers) were raised on the island.[95] Their ultimate aim in this attack was to elicit money from Government (and, no doubt, congregations in England) for the establishment of a school in the 'Black Town' where many poor military families lived. The Society asserted that the distance from the Fort at which many of the families lived was a barrier to the children's development as it prevented them from regularly attending the Charity School. In order to support the argument that a proper 'Christian' education was necessary for such children, the Society attacked the children's mothers as being a bad influence. Missionaries described the women as 'profligate' and as '... more likely to corrupt than to improve those who are near her'.[96] The letter's *coup de grace*, however, came with its assurance that there were '... instances of such mothers bringing up their children as Mahomedan and others may be considered as devoting them from their earliest years to prostitution'.[97] The link made between the 'heathen' practices of the mothers and prostitution was inescapably damning. Not only were these mothers condemned along religious lines, but to this was added the allegation (for which no evidence was provided) that such women were of 'loose morals' and were prostitutes.

After missionaries were allowed in to British India in 1813,[98] these missionary attacks on the moral character of Indian and Eurasian wives and children became more frequent. In his 1824 observations, Mr Hill, a missionary from the London Missionary Society, reiterated these assertions. He began a letter to his London congregation by complaining that he and his wife had left their small group of respectable, Christian friends in Calcutta for the 'wilderness of idolatry and sinful abomination' that was Berhampore.[99] Here, he noted, there were great numbers of 'Feringes [sic]'. These, he explained with disgust, were the children of Dutch and Portuguese fathers by native women. He noted that they were '... quite equal to [Indians] in theft and chicanery and distinguished from them by their aping of European manners, mode of dress and a smattering of English'.[100]

The result of these continued assaults on Indian companions and children was to present a picture of the wives and mistresses of soldiers as immoral, corrupting forces. From the early nineteenth century, missionaries accomplished this by employing cultural and ethnic descriptions that stressed the 'foreignness' of the women and children's behaviour to their European audiences. It is ironic, but perhaps unsurprising, that such pressure would eventually force these monogamous relationships to yield to more fleeting affairs between European men and women deemed to be prostitutes. These more temporary liaisons gained popularity, especially among soldiers who, with a pay of a shilling a day and numerous hurdles standing in the way of further assistance, were increasingly unable to afford either a companion or family.[101]

The re-construction of the 'prostitute'

Just as longer-term relationships between Indian women and European men were transformed by pressure from the colonial state, so too was the category of 'prostitute'. Here again we find that the state had a vested interest in managing the portrayal of the moral and social position of the women. In this representation, they re-wrote the category of 'prostitute' and forced such women into a legal category which they had never before inhabited – that of the 'criminal'. The 'public women' of India in the eighteenth century did not provoke the same vituperative moralising condemnation that accompanied their descriptions in Europe. Instead of assigning these women to the criminal category of 'moral degenerates' as was common in Britain,[102] Nathaniel Halhed's 1776 *Code of Gentoo Laws* suggested that such women were accepted as an important part of the functioning framework of society. The *Code* was more concerned with punishing the man who had neglected to pay a prostitute for her services, rather than punishing either party for the sexual act committed.[103] As the *Code* is a Brahminic text, care must be taken to avoid generalisations about its claims to represent the whole of Indian society; however, what is important for the purposes of this argument is that in pre-colonial India, prostitution may have at worst been viewed as a sin but never a criminal offence.[104]

The colonial state was content to re-write traditions and customs in Indian society which it perceived to be a threat, or which sat uncomfortably with the supposed 'morals' of the Company.[105] Prostitution underwent a similar transformation at the hands of the state. There was certainly a cultural gap in the understandings of 'prostitution' between India and Europe. For example, there was no equivalent of temple dancing girls in Europe to which observers could compare Indian practices. This disparity allowed medical and commanding officers to manipulate the definition of the category to suit their own needs as they translated these varied practices. British administrators, in their assertion that prostitutes could be morally and legally condemned and classed with other groups deemed to be 'criminal', were responsible for exactly such an alteration in re-working the fundamental perception of both the history and contemporary understandings of prostitution in India. The reasons that motivated this transformation were complex. As administrators sought to discourage soldiers from engaging in long-term relationships, they looked to prostitutes to satisfy the 'carnal' needs of the men. Yet perhaps equally aware of increasing rates of venereal admissions among the men, they sought to impose a heavy measure of control on the women in an attempt to curb disease. This control took the form of the lock hospital system. The lock hospital embodied the idea that low-class prostitutes, seen as being the primary transmitters of venereal diseases, should be removed and segregated from society – 'regulated' in order to affect their 'cure' and, in so doing, reduce the threat they represented to European soldiers.

The creation of a separate, criminalised, 'prostitute' category (which later contributed to the myth of a separate, historical 'prostitute caste' in India) was vital to administrators' attempts to justify the socially and physically invasive methods of venereal disease control that they proposed. Setting such women apart from the rest of Indian society enabled commanding officers and surgeons to implement a series of measures that would have otherwise provoked widespread protest. As will be discussed in the next chapter, military demands to apply regulation to greater numbers of women arose in response to the failure of early venereal disease regulation to reduce levels of disease among the troops.

After 1810, pro-regulationists argued that a key reason for the failure of the lock hospital system was the inability of surgeons and commanding officers to interfere with certain groups of women who they suggested were a potential threat to the health of the troops. These included women such as concubines, temple dancing girls, performing troupes and *nautchees,* women participating in what were, initially, distinct practices. Many of the 'entertainer' communities were nomadic or semi-nomadic. As such, they were usually positioned on the margins of society.[106] In the early nineteenth century, European observers and officials pushed this position further still from 'respectability.' Portrayed as individuals who were already irrevocably and historically degraded, who Siren-like, lured hapless soldiers and sailors into their clutches only to release them again with syphilis or gonorrhoea, the more nebulous category of 'prostitute' was easier to target and control.

Many of those deemed by European observers to be prostitutes bore little resemblance to European conceptions of 'public women'. It is no coincidence that the transformation of these diverse practices into the more common classification of 'public women' corresponded with the period of increasing political and military conquest of the British. One of the first groups of women that administrators and observers placed into this re-defined category were courtesans. Multiple structures of marriage and co-habitation operated in pre-colonial India.[107] These included relationships with both courtesans and concubines.

In the Mughal period, the courtesan was valued not just for her beauty but, perhaps more importantly, for her skills in such arts as music and dancing. In courtly capitals like Lucknow, courtesans lived and entertained in salons renowned for being centres of music and culture.[108] A wealthy patron usually maintained such women (and their households) and paid not only for the woman's quarters but for outfitting her household in his preferred style.[109] The courtesans upheld cultural and artistic traditions while also providing sexual services to a select and carefully chosen clientele. The sexual element was, however, only a small part of the women's overall role.

This position began to change in the eighteenth century. As the Company pushed forward territorially, it severely disrupted many of the courtly centres that had supported courtesan culture. Certain Company measures forced changes on North Indian elites in the early nineteenth century that in turn had a broader impact on the area's economy and political structure.[110] These included changes to revenue collection, the pensioning off of a large number of princely families, the reduced availability of 'extra' or supplementary income for rulers (including transit duties, cesses and gifts) and, finally, the disruption caused to military and mercenary families who had previously served across North India. With changing political patterns and increased urbanisation, kings and aristocrats could often no longer afford to maintain their courts. The economic (and biological) crises across North India in the 1830s further disrupted aristocratic spending. The decade saw economic stagnation, the immobilisation of capital and a series of epidemics that provoked the movement of labourers and artisans (among others).[111] The period that followed the annexation of Oudh and the exile of the Nawab in the mid-nineteenth century brought further changes for the Lucknow courtesans in particular. While the initial loss of court patronage was compensated somewhat by the new noble families and elites who continued to support them, the women no longer lived in the royal households.[112] Gradually, the social and cultural functions of their roles were stripped away. Faced with the threat of poverty, many of the women refocused their skills on the sexual elements of their roles.[113]

Like the courtesan tradition, temple dedication was not uniformly practiced across India and its assorted variations were referred to by different names. However, these practices were thrust together under the title '*devadasi*' by nineteenth-century British observers.[114] Dedicated to the temple at a young age, a girl was trained in traditional song and dance in order to perform temple rituals and religious rites. At puberty, she was symbolically 'married' to the deity. Although her main duties entailed participation in temple dances, she was also responsible for food preparation, arranging garlands for the deity, sweeping and cleaning the temple and bearing lamps to accompany processions.[115] Temple dancing girls were permitted sexual relations with the first-born sons of elite Brahmin families. However, these relationships, like those of courtesans, were carefully regulated. Any violation of this, for example, any dealings with 'lower caste' men, could lead to excommunication from both the community and temple services.[116] A temple dancing girl's sexual partner was chosen by arrangement, with her mother or grandmother wielding strong veto power over the decision.[117] This patronage served two important functions: it brought prestige to the Brahmin family and provided a source of funding for the temple. In South India, in particular, the combination of religious ritual and sex in the person of the temple dancer meant that she embodied prosperity and auspiciousness.[118] Prior to her inclusion in the common 'prostitute' category, the temple dancing girl was seen as performing a religious function, and it was rare that British observers implied any sexual nature when describing the role. However, this gradually changed in the early nineteenth century. The Company (and later Crown) attack on the *inam* grants to temples had a direct impact on their livelihood as the state increasingly interfered with and 'rationalised' the cultural and economic resources of the temple.[119] As the confidence of the Company grew, military and economic imperatives guided its approach to *devadasi*, as they did in other areas. There was no equivalent of the temple dancing tradition in Christianity; therefore, it is possible that missionaries (having nothing to compare it to) assumed the practice was obscene and immoral.[120]

The remarks such as those made in 1807 by Francis Buchanan, surgeon and Superintendent of the Calcutta Botanical Garden, reinforced the foundation for temple dancers to be considered as prostitutes (and, moreover, as a separate 'hereditary caste'). Buchanan noted that the women and their musicians

> ... form a separate kind of cast [sic]; and a certain number of them are attached to every temple of any consequence ... all the handsome

> girls are instructed to dance and sing, and *all are prostitutes*, at least to the Brahmins.[121]

While the temple dancers served a more religious function, the *nautch* girls were those called upon to sing and dance at public events as well as private functions. In his meticulously kept account book, 'Cantoo Baboo' – Krishna Kanta Nandy – *banian*[122] for both Warren Hastings and Francis Sykes, noted that in the spring of 1789, Lokenath, Kantababu's son, attended a *nautch* organised by Balgovind Das Babu to celebrate a marriage.[123] It is possible that European observers conflated the two practices because of their overlapping skills and the fact that in many areas, temple dancers played a part in pubic ceremonies, including weddings and birthdays, due to the belief in their auspicious nature.

In 1766, John Grose's *A Voyage to the East Indies* linked the temple dancing girls with *nautch* dancers and identified both as 'prostitutes'. Grose, a writer for the Company, arrived in India in 1749 and served in Bombay until 1753, when he was dismissed from the service and sent back to England.[124] Tellingly (and crucially for later conflations of temple dedication and *nautch* traditions), Grose's description of the Mughal emperor's court dancers is joined to one describing women dedicated to temples. He reduces both groups to 'prostitutes', sardonically noting that the Mughal Emperor's dancing girls made 'vows of unchastity which they religiously keep'.[125] He went further still, describing the temple dancing girls as

> [appropriated to] the use of the Brahmin-priests ... they live in a band or community, under the direction of some super-annuated female of the same profession, under whom they receive a training as regular as in an academy, or like horses in a manage, and learn all the paces, and acts of pleasing, in which they are too successful.[126]

This conjoining of categories was encouraged by the observations of missionaries. In a similar manner to how the Indian and Eurasian wives and mistresses of soldiers were portrayed as deeply immoral, missionaries could further shock their congregants through depictions of the 'degraded' character of India (and Hinduism specifically) as demonstrated by such traditions as temple dedication and *nautch* dances. Abbé Dubois, writing shortly after missionaries were allowed entry to India in 1813, catalogued many of the supposed 'vices' of India in his observations. He assured his readers that it was not uncommon to see 'sacred temples converted into mere brothels'.[127] Dubois also sought to stress that however 'upright' and decent Hindus appeared, their inner selves masked a wealth of shameful and degrading moral characteristics.[128]

British fascination with the *nautch* continued throughout the nineteenth century. *Nautch* girls were portrayed by early (usually male) European observers as tantalising, graceful and, most importantly, erotic. However, later European women attendees to the dances, like Mrs Fay in 1817, more often expressed their deep disappointment with the performances. In her memoirs, she declared that she was profoundly bored by the formality of the *nautch* she had witnessed.[129] She judged the dancing in India to be less indecent (and apparently, therefore, less exciting) than what she had witnessed in Cairo.[130] She seemed dismayed that she could detect nothing immoral or immodest in their dance. Although Fay repeated the assertion that the women were 'prostitutes', this was clearly an assumption not grounded in any personal experience. Gradually, those observations which at once stressed the strangeness of the dance and the women's (overly) modest dress while pronouncing the women prostitutes supplanted those which praised the women's abilities and grace.

Observations such as Fay's contributed to the perception that Indian women (and Indians more broadly) presented an image of respectability which masked a deeply immoral inner core. Sara Suleri has argued that the anxiety that Europeans felt about the stability of their position in India translated into narratives which stressed the 'unreadability' of the subcontinent.[131] This 'unknowability' and underlying fear is clear in those European writings on temple dancing girls and *nautchees* that highlighted the *strangeness* of Indian traditions as well as the supposed moral duplicity of the people. In her 1835 observations on the *nautchees,* Emma Roberts (while making no mention of any sexual role) was dismissive of the women's talents and disparagingly pronounced that, '[t]he dancing is even more strange, and less interesting than the music'.[132] Reverend [later Bishop] Heber uses an extract from his wife's diary to describe the *nautch* she had been invited to by the wealthy Baboo Rouplall Mullich in honour of the opening of his new house on the Chitpore Road in Calcutta.[133] While Mrs Heber was informed that the dances conveyed stories that could be deduced from the dancers' expressions and movements, she was at a loss to decipher any of these. She remained, she claimed, at once confused and bored by the dance.[134]

John Grose's 1766 fusion of the temple dancing and *nautch* traditions was one that would be frequently repeated over the nineteenth century. When it came to sexual contact with European soldiers, and accordingly (in the eyes of commanding officers and surgeons) the threat of venereal diseases, most military and medical officers at the turn of the century acknowledged that the temple dancing girls and *nautchees* represented only a very small risk to the men. Despite the strongly coercive and invasive elements contained in the early lock hospital regulations, there is strong evidence that officials initially recognised the distinction between temple dancers and the women they considered to be 'common' prostitutes. The 1805 General Order which established the lock hospital system was explicit in warning that police and Commanding Officers were not to 'interfere with the establishments attached to any of the places of Hindoo Worship'[135] and to '... observe the utmost caution not to offend the Prejudices of the Natives ... by an over-zealous exertion of the powers of this important trust, which should only be called forth by the most unequivocal truths'.[136] However, as the lock hospital system spread and the expected numbers of women presenting themselves voluntarily for inspections did not materialise, this policy of restraint gradually gave way to suggestions that temple dancing girls and courtesans should not be exempt from examination.[137]

As Chapter 2 argues, the failure of the lock hospital system led to sharper, more aggressive calls for its expansion. By around 1810, surgeons and commanding officers joined in the attacks on the two groups of women, usually describing them jointly as the 'Pagoda Establishments of Dancing Girls'.[138] This amalgamation of two groups previously acknowledged to be separate was closely linked to a suggestion of their newly defined status as prostitutes. It is unlikely that either temple dancing girls or courtesans would have associated with European soldiers, as the expense of maintaining them would have been beyond the capabilities of the average soldier. Yet, military and medical officers repeatedly targeted them after 1810. In a number of instances, commanding officers ascribed the blame for venereal disease to the courtesans, who, in most areas, lay outside the boundaries of the early venereal regulations. In 1821, Mr Dickson, the Superintending Surgeon of the Saugor[139] Field Force, in his attempt to include the women under the lock hospital's remit, angrily noted his belief that the courtesans were a 'nest' of venereal infection.[140] Such resentment by administrators continued to grow and, in the process, understandings of the courtesan as an artist were replaced by the perception of her as little more than a 'prostitute'.

The strong resentment and suspicion which many medical and commanding officers bore towards temple dancing girls makes it difficult to say with any certainty that police and commanding officers did not unofficially target the women for inspection and detention. While ultimately, military and medical officers were *officially* unsuccessful in their efforts to include the temple dancing girls under lock hospital regulations, their continued assertion that the women were 'prostitutes', combined with the similar affirmations made by missionaries, meant that in time the colonial state embraced this description.

The Anglo-Indian judiciary also encouraged changes in the perception of temple dancing girls.[141] In 1819 laws were implemented which targeted 'evil-disposed persons, chiefly women', to prevent such individuals from luring wives and daughters away to become prostitutes.[142] While this legislation did not explicitly condemn prostitution itself, it threw up barriers around those supposedly associated with it. The increasing number of laws and moral condemnations that would target such women resulted in both the expansion and formalisation of European understandings of the 'Indian prostitute'. Women who subsequently fell into this broad definition were now effectively separated from 'normal', upright society. Any female sexual behaviour outside of marriage, such as that practiced by temple dancing girls or courtesans, was increasingly in legal findings identified as 'unchaste' and 'immoral'.[143] Moreover, as has been argued elsewhere, the nineteenth century saw the re-negotiation and rigidification of the concept of 'Hindu marriage' as not only the colonial state, but so-called Hindu 'traditionalists' and 'reformers' selectively appropriated scriptures and customs to form a singular notion of 'marriage'.[144] Similarly, as Kunal Parker has argued, the criminalisation of women as 'prostitutes' was intricately linked to the formalisation of Brahminical practices which sought to uphold the primacy of patriarchy and the male head of household and firmly establish boundaries around marriage and 'appropriate' sexuality.[145]

Following further attacks on dancing girls, a number of missionaries in Calcutta banded together in the 1830s to launch an offensive on the *nautch*. In 1837, Mr Lacroix of the London Missionary Society proudly described the effects of the campaign as follows:

> The disgraceful exhibition of prostitutes dancing before an idol, which the wealthier natives adopted in order to attract European guests to the presence of the images, has suddenly disappeared. Nautches (dances) were exhibited the week before last, in only two houses ... More than fifty of the most opulent Natives, whose mansions used to be thrown open to these occasions, & thronged with European ladies & gentlemen, have this year been closed to all but the Native community[146]

Lacroix's statement raises a number of important points. First, he describes the two groups of women *jointly*, dismissing them as 'disgraceful' 'prostitutes'. Second, his statement draws our attention to the growing social divide between Europeans and Indians, even among those elites who had formerly enjoyed entertainments such as *nautches* together. Finally, his description highlights the sharp decline in the patronage of art forms such as the *nautch* in the 1830s, as groups such as the missionaries protectively drew a 'moral' cordon around them by seeking to separate them from the rest of society.

Medical, missionary and judicial campaigns against both temple dancing girls and *nautchees* were enormously successful. Later 'purity' campaigns in the 1890s were organised on an 'anti-nautch' platform which attracted campaigners such as Brahmo Samaj reformer Keshub Chunder Sen.[147] Sen bitterly attacked the women, condemning them in language suggesting them worthy of a special place in a lower circle of Dante's hell.

> Apparently a sweet damsel, a charming figure. But beneath that beautiful exterior dwells – what? Infernal ferocity. Hell is in her eyes. In her breast is a vast ocean of poison. Round her comely waist dwell the furies of hell. Her hands are brandishing unseen daggers ever ready to strike unwary or wilful victims that fall in her way. Her blandishments are India's ruin. Alas! Her smile is India's death.[148]

Through wrathful statements such as this, it is clear that the categorisation and condemnation of certain Indian women begun by surgeons and missionaries in the early nineteenth century for 'medical' and 'moral' purposes had been internalised and, perhaps more importantly, publicised as truth by British *and* significant sections of Indian society by the end of the nineteenth century. As a number of authors have argued, the complex 'derivative' discourse which Indian nationalists produced in the later nineteenth and early twentieth century set out re-defining domestic space (and with it, concepts of morality/respectability).[149] This involved a redefinition of certain groups of women into disreputable categories (by virtue of their active sexuality) that owed much to the processes begun by Company officials.

Conclusion

The combined might of these groups effectively served to cement the fate of courtesans, temple dancing girls and *nautchees* just as it did that of the Indian and Eurasian companions of European men. These changes were neither inevitable nor natural, but were shaped by the fiscal, political and medical anxieties that surrounded the Company state. As the Company expanded commercially and territorially, its armies' centrality vis-à-vis decision-making was embedded even further. While sepoys represented the majority of its fighting might, the perceived importance of a 'backbone' of European soldiers grew.

Despite the sustained attacks made on the Indian and Eurasian wives and mistresses of European men by government and missionaries alike, these relationships continued throughout the nineteenth century. An 1860 Minute by Governor-General Canning revealed that the population of Eurasian children at that time was still continuing to increase.[150] However, the move away from official toleration of such relationships, which began in the late eighteenth century, accelerated as budget-conscious officials sought to avoid any 'unnecessary' expenditure. The Indian and Eurasian wives and children of European soldiers came to be represented as a potentially enormous cost by such Company officials. As such, the Company began to discourage such lasting relationships in favour of short-term ones.

However, the women who were deemed 'prostitutes' were seen as a different kind of threat by the Company's commanding officers and surgeons. While relying heavily on these women to satisfy the perceived sexual needs of the men, officials were also keenly aware of rising levels of venereal disease among the troops. Medical officers deduced from this rise that the men had increasingly taken to visiting prostitutes whom they reasoned were the source of infection. The idea that venereal diseases emanated from the Indian woman supplanted previous ideas of two-way disease transmission. Until the early twentieth century, the bacterium that causes syphilis remained unidentified. Early theories of venereal transmission contained some odd (to modern readers) suggestions, but it was only in the eighteenth century that women were targeted as the foci of the disease. Indeed, in 1805 medical officers suggested that it was the '... unfortunate women who communicate this bitter scourge of unlawful embraces'.[151] This strengthened their belief in the need to regulate these women's bodies. Surgeons could monitor those women who fell within the remit of the lock hospital system on a regular basis. However, such un-registered women as 'illicit' prostitutes or those (like the courtesans, temple dancing girls and *nautchees*) that the regulations did not cover were increasingly resented and seen as threats by commanding officers and surgeons alike. While these latter groups of women did not fit into European understandings of 'prostitution', observers, having nothing to compare their practices to, began to paint them as 'immoral' and 'unchaste'. These two transformations affected Indian women's lives immeasurably. The demands and fears of Company officials and evangelical missionaries motivated both. What resulted was a radical change not only in the social and sexual relationships between European men and Indian women but, more broadly, in ideas about appropriate gender roles across the subcontinent.

The re-categorisation of various groups of women who would come to be designated simply as 'prostitutes' by European officials and observers proved incredibly long-lasting and far-reaching. The incremental changes made over the first half of the nineteenth century to the legal conception of the women resulted in their being judged by criminal rather than civil law when the Indian Penal Code of 1860 was enacted.[152] This categorisation would continue to be utilised by officials attempting health (or social) control measures throughout the century.

2
Regulating the Body: Experiments in Venereal Disease Control, 1797–1831

Each European soldier represented a considerable expense for the Company: one estimate suggested each man, taking into account recruitment, equipment and pensions, cost £100 on average per year to maintain in India.[1] When seen in the context of the fraught state of the Company's finances in the early nineteenth century, this cost was a sizeable investment and accordingly, one that the Company sought to safeguard. Any time spent outside active duty represented a loss to the Company: this included the time that men spent in hospital with ailments which ranged from fevers, dysentery and venereal diseases to wounds sustained from tiger attacks.[2] Of these, venereal disease was perhaps the most resented, as it was seen to arise from the sexual carelessness of the men. In addition, treatment was time-consuming and potentially dangerous and surgeons saw syphilis as having the ability to make the men more vulnerable to other ailments.

Average admissions (as a percentage of military strength) among European soldiers for venereal disease fluctuated over the course of the century and from regiment to regiment. In some cases, the rate was as low as 5 per cent (this being the average across all Madras troops in 1802),[3] whereas at certain stations, it was as high as 78 per cent (as was the case in HM's 12th Foot, stationed at Trichinopoly[4] in 1805).[5] The average rate of venereal admissions among European troops in Madras for the period from 1802 to 1835 was just under 24 per cent.[6] For cases of primary syphilis, treatment (usually using some form of mercury, as discussed in Chapter 5) took four to six weeks. When combined, these figures represent a staggering cost to the Company. Moreover, this became an issue of military security, and of the Company's ability to protect its political and commercial interests. If we assume that on average, 24 per cent of European troops were incapacitated at any given point with venereal disease and that (conservatively) their treatment took four weeks, this translated to an enormous amount of pay 'wasted' on men unable to serve. It was not simply this expense that concerned Company accountants but also the cost involved in the men's treatment (including medicines, bedding and food). Perhaps most importantly, this extended period of hospitalisation (and the sheer volume of men undergoing treatment) posed a fundamental threat to the Company's stability and security. Just under one-quarter of its European troops were unavailable to fight at any given time, due to what was seen as a preventable, 'man-made' ailment. This reduced man-power was especially significant given the Company's absolute reliance on its armies to maintain control, not just over the territories it administered directly, but over its 'subsidiary allies'.[7] The combination of costly wars of expansion or 'pacification' and rising administrative costs resulted in massive deficits. When seen in this light, the expense of a venereally-stricken army was one which the Company could ill afford.

The solution which grew up in the late eighteenth century combined the informal '*lal bazaar*' system with that of the lock hospital. The term '*lal bazaar*' denoted the area of a cantonment bazaar dedicated to regimental (regulated) prostitutes. This entailed the regular monitoring and control of women deemed to be 'prostitutes' to ensure that they were free from venereal disease. Any woman found to be suffering from the disease would then be sent to the nearest lock hospital – a hospital-cum-prison specifically established for female venereal disease sufferers, where she would be held and treated deemed 'cured'. This system was fraught with difficulties. Unsurprisingly, women resented such heavy-handed methods of surveillance and control as well as the invasive (and often dangerous) cures prescribed for their treatment. As a result, they avoided hospitalisation at all costs, much to the dismay of over-zealous surgeons and commanding officers. To borrow from James Scott, the women's 'petty acts of resistance' included everything from evading weekly examinations to fleeing the catchment area upon detecting any signs of illness.[8] The tensions that resulted from the women's non-compliance led surgeons to blame them for the failure of the hospitals.

This chapter examines the ways in which the lock hospital system developed. In the early years of its operation, the system functioned unevenly. Opportunistic surgeons and commanding officers often shaped the system in surprising ways. These officers, desperate for a solution to the problem of venereal disease amongst the troops but hamstrung by a number of constraints (or, at least, *perceived* constraints), constructed the lock hospital system in a piecemeal fashion to suit these needs.

The system, then, grew not out of any unified discourse on the best means to treat venereal disease, but from a panicked dialogue among surgeons and commanding officers. These officers were fearful of both an army weakened by disease and of provoking discord amongst the men through invasive examinations and treatments. They were keenly aware of the potential costs (fiscal and political) of both eventualities, so suggested what they saw as the only feasible and politic option – the enforced monitoring and treatment of women deemed to 'prostitutes', who would 'safely' satisfy the sexual needs of the men. According to this logic, the lengthy (often dangerous) treatment required for venereal disease would be borne by the women, rather than the soldiery, meaning fewer man-days lost for the Company.

What follows continues to explore how Company officers and surgeons manipulated names and categories to serve their needs. Regardless of the women's actual position, it was in the interests of the system's supporters to portray them as existing on the fringe of society. If, as we saw earlier, soldiers were 'ruffians', the women in and around the cantonments increasingly became 'prostitutes'. This labelling was an essential part of the process of their peripheralisation. Such marginalised women were a 'soft' target – not only were they believed to be easier to control than the men, but their treatment cost considerably less (they were, after all, 'native' women and the cost of everything – from their diet to bedding, cost less than that 'required' by a European soldier). Moreover, their position on the periphery of respectable society ensured that their harsh treatment at the hands of the bazaar police, as well as medical and military authorities, did not provoke widespread protest.

Of course, this plan was deeply flawed and cracks began to appear from a very early date. This chapter suggests that the difficulties it experienced along the way prompted the creation of a new European medical discourse on venereal disease. The failure of the system also saw further distinctions applied to the women under its remit which stressed cultural, social and racial differences. Exploring the evolution of the lock hospital system is critical to fully understand the social and political implications of its operation. In analysing the tensions which grew out of the system, we can see the ways in which ideas (and generalisations) about Indian moral, physical and cultural 'differences' developed.

This chapter will also argue that the medical intrusions pursued by the imperial state during the late nineteenth century plague epidemic had clear historical precedents in the previous forays into the lives of imperial subjects who represented so-called security threats. Thus, while Chandavarkar's argument that the actions of the colonial state in

Bombay during the 1897–1902 epidemic marked a 'new era' of intervention,[9] the invasive instruments of rule had been finely honed through years of experimentation with the lock hospital system. In the early nineteenth century, surgeons and commanding officers conceived this threat as the menace of venereal disease which stalked the men of the army. The propriety of deploying 'medical police' into women's homes for their forced examination and detention in the lock hospital was rarely ever questioned, except on the grounds of expense. While the invasive meddling into the lives of women deemed 'prostitutes' might not have prompted the same cries of protest as heard during the plague panic, the move toward an increasing ease of intervention is what is critical here. Not only were the ways in which the state and officials categorised such women (as 'prostitutes' and therefore separate from what it deemed respectable society, and as 'criminals', thereby rationalising the harsh methods employed in their treatment) important precursors to the methods later used in plague control, but the normalisation of such forms of intervention was crucial in establishing the system as *the* template for controlling venereal disease.

After a relatively short period of official operation, surgeons and commanding officers formed a more definitive vision of the women affected by the lock hospital system. As Chapter 1 suggests, faced with the need to justify invasive methods of disease control which targeted the native population the East India Company had once been so wary of upsetting, it was initially important for commanding officers and surgeons to set the women affected by the regulatory system apart from the rest of the population. The initial frustrations of surgeons and commanding officers turned more sharply towards medically-backed constructions of Indian society, employed to defend their calls for an increasingly invasive system of regulation. As time progressed and surgeons and administrators became bolder in their assertions, they attempted to broaden the category of women targeted by the regulatory system. By encouraging the adoption of European value judgements on the body of the 'prostitute', pro-regulationists were able to portray the women in much more derisory ways. Forced to justify a potentially very costly system, surgeons would increasingly argue that the main weakness in the system was the women, as prostitutes, themselves.

Medical conceptions of venereal disease

Detection of venereal disease and the difficulty of differentiating between syphilis and gonorrhoea, presented a major obstacle to effective disease control for much of the nineteenth century. It was not until 1879 that scientists formally established a distinction between gonorrhoea and syphilis. Perhaps more importantly, it was only in 1905 that Fritz Schaudinn and Erich Hoffman identified the syphilis bacterium, *treponema pallidum*. It took a further four years until Sahachiro Hata's discovery of the anti-syphilitic properties of Arsphenamine, an arsenic-based compound, were developed into the commercial cure, Salvarsan. Throughout the eighteenth and nineteenth centuries therefore, treatments for venereal infections were based on only the most uncertain medical understandings and included mercury, magnesia, sarsaparilla and potassium-arsenic compounds, among other (potentially lethal) cures. Mercury, the most popular of these for much of the first half of the nineteenth century, brought with it unfortunate side effects which could include heavy salivation, a loosening of the teeth, mental changes, kidney damage and diarrhoea.

Both European and Indian doctors used mercury, either on its own or in combination with other ingredients such as arsenic, to treat venereal diseases. Until the nineteenth century, European views on disease were not radically different from those of their Indian counterparts. Both believed that a combination of pre-existing and 'exciting' factors caused disease, in the ways these impacted and interacted with the body. Both Indian and European medical traditions saw an individual's *moral* conduct as a key factor that could indicate their predisposition to disease.[10] Moreover, for most of the eighteenth century, information-sharing between British medical officers and Indian practitioners most often took the form of dependence of the former upon the latter, as British medical officers struggled to cope with the range of diseases afflicting European soldiers in India.[11] Indian medical collaborators were essential to the smooth functioning of the military's medical service, but, after the late eighteenth century, these men were effectively barred from reaching the highest positions themselves, a restriction which lasted into the twentieth century. Until the 1830s, the Company had supported the ongoing teaching of both ayurvedic and unani medical practices through its patronage of institutions such as the Calcutta Sanskrit College and the Calcutta Madrassa, where they were taught alongside translated Western medical texts.[12] During the period when British officials such as Nathaniel Halhed and William Jones were translating Indian legal texts, there was a corresponding movement to translate Indian medical and scientific works. Similar to the debate over legal texts, it is unclear whether this move was due to 'pure' Orientalist or scientific interest in the texts themselves, or to a growing resentment of the necessary

European dependence on *vaidyas* and *hakims* (practitioners of ayurveda and unani-tibb, respectively), is debateable.

The late eighteenth century also witnessed a changed approach to European theories of disease transmission. Whereas later theories posited that differences between groups of people were inherent and unchanging, earlier climactic theories argued that environmental factors in its surrounds defined a group's characteristics. From this followed the belief that, should a group move locations, its traits would similarly change. However, as various ailments continued to strike down Europeans in India, this belief gave way to one that posited that India's climate was fundamentally incompatible with the European constitution.[13]

The early distinction made between 'natural' and 'preventable' diseases in foreign climes was also essential to later constructions, not only of disease, but of race, in India. Out of this divide, two distinct dialogues on disease arose among serving surgeons (both of which served to pathologise India as a space). Men of science believed that those diseases 'chronic' to India, such as fevers, cholera and dysentery were almost unavoidable. As has been discussed elsewhere, the debate on 'tropical' medicine grew out of understandings of such diseases.[14] On the other hand, 'preventable' ailments which were not necessarily 'located' as being unique or indigenous to India, such as venereal disease or the health problems associated with intemperance, were nevertheless believed to take on a more aggressive and serious character there.

As these new ideas gained in popularity, theories about the transmission of venereal disease also began to change. Initial beliefs about two-way transmission were now subordinated in favour of those which supported one-way transmission – often portrayed as travelling from the 'corrupted', low-class prostitute to the 'naïve' British soldier. Speaking about gonorrhoea in 1810, London surgeon Dr John Abernethy[15] observed, 'Sores ... very frequently succeed to gonorrhoea in the lower class of females, who pay little attention to cleanliness, and do not abstain from sexual intercourse.'[16] This class-based view of venereal disease was widespread amongst European surgeons in India. To compound this, venereal diseases attracted a moral opprobrium and were increasingly portrayed as yet another mark of Indian savagery or primitiveness. The nineteenth century witnessed the increasing tendency to attach words such as 'evil', 'wicked' and 'poison' to descriptions of venereal disease and its female sufferers. The supposedly elevated danger of venereal diseases in India fed later representations of the country; these served as an example of the dangers it held and the moral corruption of the people.

But why, one might ask, did officials do so little to prevent sexual intercourse between European men and supposedly 'wicked' Indian women? As alluded to in the previous chapter, officials held certain notions about the men's class and sexual 'needs'. In addition, the threat of homosexuality and masturbation aroused considerable panic. Medical men believed that both, being 'unnatural' actions, had the effect of weakening a soldiers' physique, thus destroying his ability to fight effectively. Masturbation was regarded as being mentally debilitating and a causal link to other diseases. In 1839, physician Michael Ryan wrote that masturbation, '... notoriously undermines the natural powers of the constitution, at every period of life, and leads to the production of the most formidable and incurable diseases.'[17] Ryan continued on to note that the '... emission of the seminal fluid without a spontaneous natural impulse, is injurious to health and weakens the slender thread of human life.'[18] Thus, medical officers reasoned, in order safely to control the men's sexuality, regular visits to prostitutes were necessary. Yet, in order to prevent an attack of the venereal 'plague' on the men after visiting such women, some form of protection was urgently needed. The widespread use of prophylactics in the form of condoms would not be adopted until the twentieth century. Instead, 'protection' would come from the knowledge that the women whom the men slept with were 'healthy'.

Early experimentation with lock hospitals and regulation

The concept of a hospital dedicated to venereal cases was not new. A lock hospital to care for venereal patients (as opposed to the 'loque' – the earlier incarnation as leper hospitals) opened on Grosvenor Street in London in 1746.[19] The fact that these venereal hospitals developed from leper institutions suggests that venereal sufferers were isolated and stigmatised in a similar manner to those afflicted with leprosy. The parallels with leprosy can also be seen in the social stigma increasingly attached to venereal disease. Josephine Butler (the champion of the anti-regulationist cause in Britain from the late 1860s), asserted in her memoirs that the regulation of prostitutes was '... first suggested by Aulas in 1762 and by Restif de la Bretonne in 1790. It was brought into full operation on the eve of the establishment of the French Empire in 1802.'[20] However, as this chapter argues, the establishment of the London lock hospital and French regulation highlight only part of the story. The particular way in which military and medical men came together to develop these two ideas in early nineteenth century

India had a hugely significant impact on both Indian society and the construction of ideas of human 'difference'. Moreover, the long years of experimentation with sexual regulation in India provided a more familiar template for later use in Britain, as well as furnishing a number of surgeons and commanding officers who had direct experience of the system of regulation.

The idea of superintending the bodies of women for the purposes of controlling disease grew up organically around cantonments in India in the 1780s. The Indian experiments with regulation in the late eighteenth century were charged with a different set of meanings than their French counterparts. In India, surgeons and officers attached more urgent imperatives to ridding the army of venereal disease. The maintenance of newly gained territory and political concessions in India was seen to rest on the strength and fighting ability of the army (and the European soldiery as a central element within it). In this light, any impediment to the army's ability to mobilise its men, such as that posed by rising levels of venereal disease among the European corps, could endanger the Company's position in India.

As early as September 1766, the military took steps to curb rising venereal infection rates among the European soldiery in India. At this time, it issued an order which outlined proposals to discourage soldiers from contracting such diseases. Yet, interestingly, unlike later actions, this order targeted the men themselves, in the hopes that the threat of punishment was the best deterrent. The order sought to establish a clear distinction between those men admitted into hospital with 'common and natural disorders' and those admitted as a result of venereal infections.[21] Venereal disease sufferers would have five rupees deducted from their pay (as opposed to the standard stoppage of three rupees).[22] Of course, with little clear understanding of how such diseases were transmitted (other than an awareness that this occurred through sexual contact or from mother to child), such a penalty failed to tell the men *how* they should avoid contracting the disease, other than the vague insistence that they refrain from visiting 'Publick Women'.

However, such punitive measures that attempted to target the European soldiers themselves were short-lived.[23] Medical officers feared that any penalty inflicted on a soldier would increase the likelihood that he would attempt to conceal his ailment and this would, in turn, lead to a more serious case. Perhaps more critically, commanding officers feared that such punishment would foment discontent in the ranks and contribute to unproductive or mutinous behaviour. As commanding officers and surgeons grew more reluctant to inflict any sort of 'punishment' on the men, alternative methods were sought in order to staunch the creeping tide of venereal infections.

In examining the ways in which such 'alternatives' were crafted, certain priorities surrounding treatment become evident. The health and fighting ability of the men was deemed to be of primary importance. Any lengthy treatment, or that which might do lasting damage to the men's health was to be scrupulously avoided. Dr Price, arguing for the establishment of a lock hospital system in 1805, was at pains to point out to government that men who had been subject to repeated attacks of venereal disease were more susceptible to fever, dysentery and 'the whole train of distempers to which Europeans in this Country are liable.'[24] He then went on to warn that they were at a massive disadvantage as their ability to resist future attacks '... when their constitutions must be impaired and their whole system relaxed from the several courses of Mercury they necessarily must have gone through.'[25] This fact, when combined with the movement away from punishing the men, either through the treatment itself or pecuniary methods, reflects the peculiar bind that surgeons and commanding officers were in. While desperate to preserve the health (and therefore fighting ability) of these European soldiers, they trod carefully when carrying out such measures, as if fearful of the men's wrath. Commanding officers assumed, by and large, that punishment would encourage the men to conceal any signs of disease and would provoke insubordinate behaviour, making the corps unmanageable.

Military officers believed that as the majority of soldiers were from the 'lower orders', they were completely unable to control their sexual urges. The solution was the *lal* bazaar and lock hospital system. Thus, an area of 'regulated', managed prostitution was combined with the lock hospital where surgeons sent 'diseased' women for treatment, to remain detained there until cured. These would serve to regulate the 'public' or 'bazaar women' and ensure that any among them suspected of being diseased, was promptly sent to the lock hospital for treatment. These hospitals did not offer the women any alternative remedy to those given to the men. The same mercurial medicines formed the basis of treatment and women's 'cures' took the same amount of time as soldier's. However, surgeons and commanding officers hoped to minimise the men's exposure to venereal disease by instead subjecting their sexual companions to monitoring and treatment. Surgeons and officers designated these women to be 'prostitutes'. This categorisation was of great political and strategic importance. Had it been acknowledged that it was the wives or mistresses of the men who were being targeted, no doubt further discord would have resulted. Categorising the women simply as 'prostitutes' (whether or not this classification was correct), automatically forced a social and emotional wedge between such women and the soldiery.

Although there is no official mention made of a lock hospital at Surat (or in Bombay more generally), the presence of an (unofficial) regulatory system is recorded as early as 1781. In that year, Samuel Hickson, an officer with the Company, described the practices of Brigadier Thomas Goddard's[26] army (in which he was serving) in a letter to his brother,

> ... so general is the intercourse with the most abandoned prostitutes ... there was a *part of the camp allotted for them, to pitch their tents in,* which went by the name of the Loll Bazar [sic] ... [and] when we arrived in Cantonment near Surat, the following remarkable orders were given out from head quarters: 'a committee of Surgeons to assemble on the ____ instant, to *examine the public women of the Bazar; those who are found disordered are to be sent to the Hospital at Surat*'.[27]

It seems probable that Goddard was not alone in experimenting with this method of disease prevention. Given the widespread support enjoyed by lock hospitals in Madras in the early nineteenth century, it is likely that the *lal bazaar* system was loosely and unofficially established across that presidency as it was in Bengal and Bombay. An 1818 argument over expenses between two surgeons in Berhampore reveals that a lock hospital had been in operation at that station since 1788.[28]

The system and hospital that Berhampore experimented with in 1788 was formalised in 1797, when official lock hospitals were ordered at that station and Cawnpore, Dinapore and Futtygurh.[29] These hospitals were located within the cantonment bazaars and surrounded by a mud wall. The physical construction of these buildings ranged from small huts to larger brick buildings. All included a guard room, wards for the inmates and external boundary wall. Both a native and European doctor were on the payroll, as was a 'woman servant' or 'matron', who, with her assistant, were responsible for looking after the women interned within.[30] The responsibility for managing the bazaar women lay with the *kotwal*, or superintendent of police for each cantonment. Surgeons established regular inspection days, however, any woman suspected of being diseased was ordered, regardless of the day, to be pointed out and sent directly to the hospital for examination. Women were forcibly detained in the hospital until they were deemed 'cured'. A woman was not permitted to return to her home without first obtaining a Certificate of Discharge, which was then delivered to the *kotwal* responsible for her bazaar, to notify him of her improved condition.[31] Even during these early years, the system was a punitive one; its methods relied on both implied and explicit threats as well as the police to operate. Such a focus reinforced the notion that the women, as 'prostitutes', should be classed as 'criminals'. Not only were they seen as dangerous transmitters of disease, but as unruly 'prostitutes' whose sexuality needed to be carefully monitored and managed so that it would not 'infect' the rest of society.

Levels of venereal disease continued to rise during the late eighteenth century and by the early nineteenth century, officers grew progressively more anxious. There was clearly an awareness of the four early lock hospitals beyond the stations themselves and across both the Madras and Bombay presidencies, as surgeons there soon petitioned to apply the system more broadly. In his 1805 report to the Medical Board, Assistant Surgeon Price of His Majesty's 12th Foot at Trichinopoly, arguing in favour of a broader, official system of regulation, claimed that the incidence of venereal cases had risen so sharply that, of a total of the 23 patients he was currently treating in hospital, 18, or 78 per cent, were venereal.[32] Admissions for venereal disease in Madras remained high throughout most of the first quarter of the century, representing approximately one in every four soldiers admitted into hospital.[33] (See Table 2.1).

Medical officers deduced from the rise reflected in Surgeon Price's report that the men had increasingly taken to visiting prostitutes, whom, they reasoned, were the source of infection. Assumptions such as this clearly reinforced the notion that the blame for disease lay squarely with the woman. The men's 'tendency' towards visiting prostitutes, if such was the case, could not have been aided by the fact that for most soldiers, drill was limited to nine hours each week.[34] While one observer noted that the men were also encouraged to engage in 'manly sports',[35] for the most part, they were free the rest of the day and often spent it roaming the cantonment and adjacent towns, much to the dismay of the army's surgeons, who saw such perambulations as the source of both disease and intemperance.

Most cantonments, while equipped with a canteen, did not provide soldiers with alternate recreational activities. For much of the century, regulated drink and sex were the only things consistently on offer for the men's 'leisure' activities. While this fact is ironic considering the great fears that commanding officers and surgeons maintained about both activities, it goes some way towards explaining why venereal

Table 2.1 Venereal disease admissions among European troops in Madras, 1802–1835

Years	Percentage of admissions	
1802	5	
1803	19	
1804	18	
1805	26	Lock hospitals established in July 1805
1806	27	
1808	45	
1809	38	Lock hospitals partially discontinued in July 1809
1810	31	Lock hospitals partially re-established in June 1810
1811	28	
1812	31	
1813	35	
1814	30	
1815	19	
1816	21	
1817	25	
1818	26	
1819	27	
1820	22	
1821	24	
1822	25	
1823	21	
1824	24	
1825	18	
1826	16	
1827	16	
1828	18	
1829	19	
1830	12	
1831	15	
1832	15	
1833	33	
1834	33	
1835	28	Lock hospitals discontinued from 1 March 1835

Source: Annual Report on the Military Lock Hospitals of the Madras Presidency for 1871 (Madras, 1871).

disease and intemperance remained two of the army's greatest health concerns. The Company was not particularly interested in providing 'healthier' alternatives for the men. Essentially, they viewed the men as mercenaries and had little desire to spend more money than they needed on the men's upkeep. Portraying the men as low-class, degraded ruffians provided a convenient excuse to pay as little attention to their intellectual (or physical) wants as possible. The Company's responses to the pleas and suggestions made to attend to the intellectual, social or sexual needs of their men are revealing; a pattern of employing the least expensive means of each quickly emerges. What was sought was very much an army 'on the cheap'. While the military and defence represented the greatest proportion of the Company's spending – these costs did not extend to such 'liberal' measures as the men's intellectual or emotional well-being.

Admittedly, a few campaigns for the 'moral improvement' of the soldiery began in earnest in the 1820s. Officers and surgeons supporting this proposed moral uplift often linked it with the health of the men, suggesting it as a more 'cost effective' way of countering the rising costs of the army medical establishment. An anonymous medical writer (calling himself 'A King's Officer') wrote his recommendations for such improvement in 1825. Reading, 'private theatricals', and savings banks were to be encouraged, as should a form of competitive gardening, whereby 'small premiums' would be awarded to those men found to be 'the most diligent and successful horticulturalists.'[36] Perhaps unsurprisingly, competitive gardening failed to take hold. It was not until the 1860s that spaces such as regimental libraries (often tiny holdings of a handful of books) became slightly more common in the cantonments.

The demand for solutions to the growing problem of venereal disease resulted in the restructuring and expansion of the *lal* bazaar and lock hospital system. In 1805, the Medical Board in Madras sanctioned further lock hospitals and with them, a more explicit system for the policing and hospitalisation of women suspected of being diseased. An 1805 General Order formalised the earlier measures, stating that the Civil Magistrate for each particular cantonment would give monthly orders regarding the 'discovery and inspection' of all women deemed to be prostitutes.[37] By this order, commanding officers and the troops themselves were required to point out any such women that they 'may have reason to consider as objects to be placed under the sanitary restrictions' to the magistrate or their commanding officer.[38] Through this order, the burden of proving one's 'innocence' or, perhaps more accurately, 'virtue' was placed on the individual woman herself.[39] However, it was hoped that all women who belonged to the nebulously defined category of 'prostitute' would voluntarily and happily present themselves to police or doctors upon first detecting any symptoms of disease and, moreover, 'bless their humane benefactors' for providing them with medical treatment.[40] It is difficult to determine if this stance was the result of a curious naïveté among those drafting the order or whether they were instead aware of the limitations and risks involved in embarking on such a method of treatment and chose to wilfully gloss over them with such artificially charitable language. In view of the fact that lock hospitals in India emerged out of the very clear desire to avoid inflicting prolonged and unpleasant treatments on the men, the latter seems more likely and surgeons clearly understood the potential protests that might arise. Yet, despite this fact, at least some of the surgeons appeared to sincerely believe that the women who were the subject of these orders would welcome the chance for treatment, invasive as it was. These surgeons had obvious faith in the curative power of western medicine and either did not entertain the notion that such restraint and treatment might be considered wholly objectionable and completely foreign to the women, or, simply chose to ignore this possibility.

However, a more realistic group of surgeons also had a hand in the drafting. The 1805 Order stipulated that the military hire an appropriate number of peons 'to preserve order and regularity' and to 'prevent the patients at all leaving the Hospital without permission of the Surgeon.'[41] It seems most likely that this force of peons was chosen from among the bazaar's native policemen; they were then placed under the direction of the *kotwal*. Peons were ordered to seek out all women believed to be diseased and hand them over to the *kotwal* or the hospital directly. The pseudo-altruistic voluntarism hoped for by the first group of surgeons was soon replaced by a greater emphasis on coercion as the numbers of women 'voluntarily' submitting themselves to the hospitals failed to materialise. The power given to surgeons to detain the women, combined with the prerogative to detect the infirm among them by means of a strong bazaar police was to be supplemented by the supposedly privileged knowledge of the 'matron' who exerted a more subtle coercive power.

The terms 'Matron' and 'Decent Widow of Cast and proper years' were euphemisms for 'Madam'. It was thought that such a woman would be best suited to the task of monitoring the women as she was familiar with the local brothels and bazaar prostitutes. She would 'manage' the women in the bazaar.[42] The matron appears to have played differing roles depending on the cantonment she operated in. While most surgeons outlined her duties as cooking, cleaning and regulating 'the Economy and Internal Management of the Hospitals,'[43] in many cantonments, her primary role was that of a vigilant detective, appointed in the expectation, as one observer later commented, that 'old poachers make the best game-keepers.'[44] In this investigative capacity, she was responsible for ensuring that the women in the bazaar were not diseased and, if she suspected any of being so, making certain that they were promptly 'escorted' to the hospital for treatment.

Surgeons and officers saw the matrons, peons, police and bazaar *kotwals* who formed the surveillance and enforcement end of the system as absolutely necessary to its successful operation. Medical officers frequently made calls for a 'well constituted and efficient *medical* police' to ensure the system's proper functioning.[45] It was expected that this force would serve two primary goals: the detection (and informal examination) of women suspected of disease, and their forced escort to the lock hospital for full examination and treatment. Surgeons reasoned that the regulatory system was markedly more successful at stations (such as Trichinopoly and Bangalore), where police were more assiduous in detecting and detaining suspect women.[46] In this manner, the punitive foundations of how 'prostitutes' were to be dealt with were laid out. Military and medical officers were careful to construct their portrayals of these women as being fundamentally different from the rest of 'respectable' Indian society. In so doing, it was easier to justify the extreme methods of treatment and punitive control that the women were subjected to at the hands of police and medical officers.

While peons and police were thought essential to monitor the women and ensure they were brought into the hospital, the hospitals themselves relied on native doctors to function. While responsibility for overseeing the hospital rested with a European surgeon (normally of the 'assistant' rank), a native doctor (or 'assistant') carried out most of the treatments. Although these practitioners played a vital role in the operation of the hospitals, very little direct mention is made of them in the records, save the expense of their salaries, or in cases where it was alleged misconduct had taken place. It is interesting to note that this had not always been the case. Early lock hospital orders acknowledged the importance of both the native doctor and local, 'country' medicines to treat the women.[47] The provision for the latter was set at two rupees, eight annas a month per patient in 1799.[48]

In a similar manner, through an examination of some of the indigenous treatments listed, it becomes clear that many were similar to those employed in Europe. While outlining the costs for the treatment of two women infected with venereal disease in Dinapore in 1799, the Head Surgeon, Alexander Carnegy lists the 'country' medicines used in treatment: quicksilver (or mercury), sulphur and opium.[49] European practitioners, of course, also relied on these remedies. The later attempts made to stress the superiority of European medical and scientific methods were, in the case of the treatments themselves, just that. When dealing with venereal disease, European surgeons in India made use of a mixture of practices – borrowing from both European and Indian methods of treatment. In fact, for most of the century, they remained reliant on 'country medicines' and 'native' assistants (although the 'native doctor' was nominally replaced). While their skills and knowledge continued to be relied upon, as the regulatory system was formalised, European surgeons, attempting to establish themselves as 'superior' to their Indian counterparts (and perhaps feeling threatened by the position of the native practitioners), now began to refer to them more dismissively as 'assistants' or 'apothecaries' in an attempt to reinforce their inferior status. Later reports to the Board of Commissioners took this posturing even further; some blamed native practitioners for the failure of lock hospitals in curbing venereal disease rates.[50]

While the native doctor was now down-graded to an 'assistant', the continued dependence on 'country' medicines remained, as can be seen in the outline of expenses for the Kaira[51] lock hospital in 1817 (See Table 2.2). Indeed, we can see from this list that European medicines do not feature at all in the administration of the Kaira lock hospital. European surgeons attempts to stratify and prioritise the knowledge

Table 2.2 Monthly cost of maintaining the Kaira lock hospital

	Rupees	Annas	Pice
Allowance to the Surgeon in charge of 2 Rs per day	60	..	..
1 Native Assistant	20	..	..
1 Matron	12	..	..
1 Do. Cook	6	..	..
1 Halalcore*	5	..	..
1 Cooley to every ten Patients @ 5 Rs per month			
Monthly allowance for the purpose of providing Country medicines, bandage cloth &c.			
An allowance of 64 Reas** per diem for the diet only of each native patient, or in scarce seasons 1 seer of rice and 40 Reas in money for the diet of each native patient			
1 Matt and Cambley*** for each patient on admission to the hospital supplied by the Commissary General			

*A sweeper, or cleaner.
**Eight Reas was equivalent to one half-anna.
*** A blanket.
Source: Establishment of a Lock Hospital at Kaira and Baroda. Extract, Bombay Military Consultations, 14 May 1817. APAC, F/4/563/13819. Bombay Military. Board of Commissioners, Collections, 1818–1819.

of European over Indian practitioners had apparently not yet reached the pharmacopeia.

The lock hospital was primarily, if not wholly, aimed at protecting the health of the European soldier, rather than the sepoy or even those of the officer class or European Civilians. Due to their ability to access greater financial resources, it is likely that most infected officers sought more discreet private medical treatment. Accordingly, the numbers of men treated in such a way does not appear in the general medical accounts.[52] While a handful of officers were hospitalised with venereal disease, this number was insignificant in comparison to the number of soldiers hospitalised for the same. More tellingly, any concern, or even awareness, about venereal diseases among either sepoys or the Indian public was almost completely lacking for most of the nineteenth century.[53] There is only minimal evidence that gives any idea as to the number of sepoys with such diseases. However, the few reports that are available suggest that among most sepoy regiments, levels of venereal disease were lower (though still significant) than that of their European counterparts (see Table 2.3).

Emerging disease theories which linked disease to race could not explain why, throughout the nineteenth (and indeed twentieth) centuries, Indian sepoys appeared to display significantly lower levels of disease than their European counterparts. Surgeons suggested a few possible explanations as to why levels of disease among both European officers and sepoys were relatively low. The most popular posited that as greater numbers of both groups of men were married, they were less likely to visit prostitutes. Alternatively, even without the presence of a wife, as the families of many sepoys lived nearby, their low disease levels (and good behaviour) were thought to be the result of family

Table 2.3 Venereal disease admissions at Kaira, 1 June to 30 November 1812

Number of men in hospital affected with venereal complaints during the month of:	European	Natives	Total
June	59	29	88
July	64	35	99
August	95	39	134
September	74	43	117
October	111	44	155
November	112	50	162

Source: Bombay Military Consultations, 31 December 1812. APAC, F/4/563/13819. Board of Commissioners, Collections, 1818–1819.

and community influence. The lack of family pressure on European soldiers was often cited as an excuse for their visits to prostitutes, or, as one superintending surgeon stated, soldiers 'habits are ill-calculated to guard against temptation.'[54] Finally, it is possible that like officers, sepoys sought treatment privately from native practitioners who left no official record of either ailment or treatment. It is entirely possible that the number of venereally infected sepoys went severely under-reported due to their reduced access to European medical assistance. However, in any event, this fact was of little concern to the Company, whose focus resolutely remained the more expensive European soldier.

Balancing the budget: the costs of regulation

With an increasing panic around the health of these soldiers, the introduction of lock hospitals across India appeared to provoke very little dissent, in sharp contrast to the explosive debates that would take place in the years that followed the introduction of the Contagious Diseases Acts in Britain. Such dialogues as there were in early nineteenth century India were usually internal to the medical and military corps and the Government of India. These centred around two main issues: the propriety of inspecting women and the expense incurred in maintaining the system (in comparison to the benefit produced). It was the latter of these two concerns that took the greatest precedence. Reports written in support of the system attempted to assuage the government's pecuniary fears. In one such report, the Madras Medical Board insisted that the establishment of the lock hospital system was '... well calculated to preserve the constitutions of the Soldiers and the consequent efficiency of the Military Service without seeming likely to incur any considerable additional expense.'[55] A report by Surgeon Patterson of HM's 25th Dragoons estimated that through the joint action of the 'efficient system of police' and the lock hospital at Bangalore, 20 to 25 of the men under his care had 'been saved annually to their Country and their services preserved to this Establishment.'[56] The Medical Board, extended Patterson's calculation further, asserting that if it was applied to all European Regiments currently serving in Madras, the annual number of men 'saved' would amount to at least 250 soldiers, who, they noted 'the value of whom in a pecuniary point of view infinitely exceeds the amount of the Expenses incurred by the whole Lock Establishment.'[57] This concern with establishing a cost-benefit analysis was a key feature for the whole of the system's operation. The military commissioned reports to measure the success of lock hospitals against financial outlays. By 1808, there were 18 lock hospitals in operation in Madras, treating just over 3,500 women annually at the approximate cost to government of Rs 7,200 for operating expenses alone.[58] While early reports from these hospitals optimistically suggested that they were successful, later statements more often drew the opposite conclusion.[59]

Account sheets, such as the one seen below for Madras, measured, in Rupees, Annas and Pice, the precise amount spent in treating the women. This was weighed against the numbers of women treated and 'cured' by each hospital, and, most importantly, the number of men admitted into the cantonment hospital for venereal disease. (See Tables 2.4, 2.5 and 2.6).[60]

While there was a slight temporary drop in the men's admissions following the introduction of lock hospitals, levels of venereal disease among European soldiers in Madras (in relation to their overall strength) remained alarmingly high. This fact may go some way towards explaining why commanding officers and surgeons in that presidency were so

Table 2.4 Lock hospital expenses for the months of April–June 1808

	Total		
	Rupees	**Annas**	**Pice**
St Thomas's Mount	30	7	40
Poonamallee	73	21	..
Arcot	82	39	..
Cuddalore (New Town)	88	36	..
Vellore	130	28	40
Wallajahbad	121	16	..
Seringapatam	65	18	..
Bangalore	362	28	40
Bedenore	21	1	40
Vizagapatam	71	16	40
Chicacole	87	6	..
Ganjam	119	44	40
Bellary	125	27	..
Gooty	98	15	40
Trichinopoly	71	1	40
Tellicherry	85	28	64
Cannanore	74	38	40
Masulipatam	83	35	40
Total (in Pagodas)	**1794**	**4**	**64**

Respectively, St Thomas' Mount, Poovirundavalli, Arcot, Kudalur, Vellore, Walajabad, Srirangapatna, Bengaluru, Bidanur, Visakhapatnam, Srikakulam, Ganjam, Bellary, Gooty, Tiruchirappalli, Thalassery, Kannur and Machilipatnam.
Source: Lock Hospitals: The Superintending Surgeon's Reports. APAC, F/4/345/8032.

Table 2.5 Lock hospital admissions, 1 January to 31 December 1808

Garrisons & Cantonments	Number admitted	Number cured	Number dead	Remain
St Thomas's Mount	33	26	1	6
Wallajahbad	194	179	3	12
Arcot	288	275	..	13
Poonamallee	38	38	..	..
Vellore	212	194	..	18
Cuddalore	74	68	..	6
Trichinopoly	131	87	..	44
Masulipatam	400	324	..	76
Vizagapatam	26	15	..	11
Ganjam	72	67	..	5
Chicacole	62	44	..	18
Seringapatam	277	67	..	210
Bangalore	1443	563	10	870
Bedenore	18	8	..	10
Bellary	100	85	3	12
Gooty	114	77	1	36
Cannanore	20	12	..	8
Total	**3502**	**2129**	**18**	**1353**

Source: Lock Hospitals: The Superintending Surgeon's Reports. APAC F/4/345/8032.

Table 2.6 Number of cases of venereal disease in European Corps, 1802–1808

	Average numerical strength	Number of cases treated	Proportion of the number of cases treated to numerical strength (per cent)
From Aug 1802	5,415	491	9
1803	5,173	1,796	34
1804	4,569	1,533	35
To July 1805	5,524	1,010	18
From Aug 1805	4,729	1,413	29
1806	6,066	2,965	48
1807	7,646	4,153	54
To July 1808	8,314	4,424	53

Source: Letter from the medical board on the increase of venereal disease in HM's Regiments, 6 November 1838. Madras Military Consultations, 30 October to 13 November 1838. APAC, P/267/15/107.

desperate to support the continuance of the system as they could see no other way of reducing these numbers. In response to similar returns in Bombay (which suggested the failure of the lock hospitals significantly to affect levels of disease among the troops), the Medical Board there raised a complex set of concerns about the system. They stressed their doubts as to the efficacy of the endeavour. This, they suggested, could be due to the difficulty surgeons and commanding officers had in confining the women. However, somewhat surprisingly, the Board also questioned the propriety of this use of force and the legality of resorting to such methods of compulsion.[61] However, it is clear from the volume of statements by surgeons that most had little compunction about the propriety of rounding up and forcibly detaining women. Instead, they simply lamented their lack of resources and actual success in doing so. Mr Hamilton, Superintending Surgeon at Chunar, is a rare example of one surgeon who appears divided regarding these methods of compulsion. However, even Hamilton grounded his reasoning in economic concerns. Writing of the increase in women with secondary symptoms of syphilis appearing at the bazaar hospital, he noted that

> The propriety of keeping those women so long in Hospital or indeed of admitting old disparate cases at all is very questionable. For the sake of humanity it may be proper but the immediate object in view is totally defeated as they remain a mere burden to the Hospital.[62]

As reports continued to reflect inconclusive results, Bombay and Bengal began, from about 1807, to gradually discontinue or reduce lock hospital operations; at the same time, they attempted to tighten up the system. This meant greater allowances for peons to detect 'wayward' or recalcitrant women. At that time, the Bengal Medical Board authorised the allocation of further funds to hire more peons.

By 1810, the Board of Commissioners was convinced that lock hospitals had not been successful enough in reducing the number of venereal cases among the troops and ordered their partial abolition. However, surgeons were successful in arguing that a few hospitals should remain in operation as an 'act of humanity' for the 'wretched females' still undergoing treatment there. This order reduced the number of hospitals in Madras from 18 to 8 and set their expenses at a fixed amount, not to exceed 3,000 Pagodas annually.[63] After the reduction of lock hospital numbers in Madras, the Medical Board hinted at possible 'loopholes', which would allow surgeons to carry on with the system, even without the presence of official, physical establishments. They directed that at both civil and military stations where no lock hospitals were established, medical officers should '... afford every medical assistance in their power to such unhappy women as may require it.'[64]

The Madras Medical Board consultations emphatically and almost desperately noted their surgeons' insistence that the lock hospitals served as the 'only means in our power' to halt the progress of venereal disease among the troops.[65] This marked the start of a concerted campaign by commanding officers and surgeons to argue for the re-establishment of the system. In an effort to reinstate the hospitals, in their reports to the Company's Leadenhall Street offices, they stressed the 'alarming extent' to which venereal disease reappeared among European troops.[66] A wave of venereal infections struck Madras in 1810, and surgeons were quick to point to this outbreak in an effort to force the Board of Commissioners to reconsider its decision. Surgeons now argued that if the Board were to more carefully examine their reports, it would observe that at stations in which 'circumstances were favourable' – i.e. stations where peons and police were given more aggressive coercive powers to capture and detain women, disease rates showed a significant decrease.[67]

In a similar manner, medical officers in Bombay, arguing for the re-establishment of a 'temporary' lock hospital at Kaira, repeated Price's earlier argument. They asserted that, given the debilitating effects of mercurial treatment, men became more susceptible to fevers and 'other disorders which often have a fatal termination.'[68] Therefore, they argued, '... any reasonable expense incurred with the view of preventing or lessening so great an evil must be in a political and humane point of view beneficially applied.'[69] Such statements laid bare the true aims of the regulatory system. The women undergoing treatment (the very same dangerous, mercurial cures) were insignificant as they were less valuable to the Company than the European soldiers. While this is no surprise given the way in which the system was constructed, this particular statement is somewhat unusual in baldly spelling out these priorities.

Surgeons and military officers presented the Board of Commissioners with only one solution to deal with the supposedly imminent pending cataclysm – the re-establishment of lock hospitals along stricter lines. These desperate proposals, pleas and threats at least temporarily convinced the Board of Commissioners of the utility of lock hospitals and in 1811, it agreed to the re-establishment of hospitals at Bangalore, Wallajabad and Poonamallee.[70]

'Martyrs to the effects of their licentiousness': morality and disease

At the same time as surgeons rallied for the continuance of the lock hospital system in India, the influence of the evangelical movement was growing in Britain. While initially focusing its moral reform efforts on the poor in Britain, as the empire expanded, so too did cries for the salvation of 'heathen' peoples around the world. In 1813, following a round of charter negotiations, the East India Company was forced to allow missionaries into British India. While not altering the desired or actual actions of the army, the moral tones cast by the evangelical movement influenced the language used by both medical and military officers in their reports on the lock hospitals. Far from unseating their support of lock hospitals however, these men simply absorbed the moralising language of missionaries to reinforce their own arguments. Descriptions of both disease and 'prostitutes' after 1813 railed in almost biblical terms of men 'saved' from the 'evils' of venereal infection and the potential benefits to humanity that its eradication would bring. These 'humanitarian' arguments never identified women individually, but instead saw them collectively as 'prostitutes' or 'camp followers'. Reports portrayed these women in two ways: either as villains, inflicting disease on the hapless, naïve soldier or as suffering victims, desperately wanting or needing the benevolent treatment that a lock hospital provided. The Madras Military Board (when seeking to stall the abolition orders for lock hospitals) made use of this 'humanitarian' argument as early as 1809. The Board insisted that the abolition of lock hospitals would 'expose the unhappy objects now under care in them to the united evils of famine and disease.'[71] In 1813 the Board repeated this newfound ethos, pointing out that such hospitals 'afford relief to a class of suffering females, who being almost exclusively dependent upon the European branch of the Army, have peculiar claims to the protection of Government.'[72] The 'uplifting' role played by lock hospitals in attending to the needs of a group of women previously described as 'wretched' thus began to be wielded as a tool to justify their continuance. Mr Mansell, Garrison Surgeon at Allahabad, reiterated this in 1821, when he described the lock hospital (which, much to his dismay, had recently been abolished), as an 'asylum' and 'retreat'. In a less sympathetic vein, however, he continued on to describe the women inmates as '... martyrs to the effects of their licentiousness'.[73]

These descriptions alternated in reports, apparently dependent on the perceived leanings of the intended audience. Military and medical boards often shifted their defence of the lock hospital from that of a military imperative to humanitarian and moral 'duty'. This argument noted that the Company needed to establish and maintain these 'benevolent' institutions to ensure that European and Indian men were protected, as much as possible, from the ravages of venereal disease.

While acknowledging the reluctance of women to 'avail themselves' of the humane treatment forced upon them, the secretary of the Bengal Medical Board wrote emphatically of the, '... great good, not merely to the unfortunate beings immediately received in them but more remotely to the health of the European and Native soldier.'[74] However, while almost romantically extolling the great benefits of these hospitals, the tone of this letter soon took a more authoritarian turn to reveal the primary goal of the lock hospital. The Board's belief that harsher methods should be adopted to prevent and detect 'runaway prostitutes' is declared in no uncertain terms. The Board sought a guarantee that magistrates and civil authorities would not allow such truants to remain in their jurisdiction. These authorities, surgeons asserted, should be *required* to assist the military in rounding up such women in the interests of promoting the 'security of the publick [sic] health.'[75] Here, the intended meaning of the term 'public health' should not be misunderstood. Military and medical officers continued to use this phrase to appeal for greater support from the government, however, their implied 'public' was not the Indian masses, but instead only those Indians whose health might negatively impact on that of the European soldiery.

An 1827 argument between civil and military surgeons in Futtyghur, reveals how the great 'humanitarian' benefits of the lock hospital were a convenient cover for their explicitly military goals and little real care accorded to the health of the Indian 'public'. Until the late 1830s, Futtyghur was a station with relatively few European troops. Despite this, medical officials ordered a lock hospital established there in 1797.[76] It appears that at this station, the hospital's original aim was to curtail 'the ravages of disease amongst the Musselman [sic] population so numerous in the city and District.'[77] Accordingly, the civil surgeon oversaw the Futtyghur lock hospital. By the 1820s, arrangements of this sort were rare, with only Almorah[78] maintaining a similar configuration; instead, the majority of lock hospitals were located only where there were large concentrations of European troops. These were supervised directly by military surgeons.[79] While the original orders had sought to redress what was a 'civil' problem, by the 1820s, any interest in improving the 'public' health through these hospitals had vanished and they were almost exclusively viewed as tools for improving military efficiency. However, the Futtyghur lock hospital remained under the supervision of the civil surgeon until 1827. At this time, Mr Webster, the Assistant Surgeon attached to the 2nd Extra Regiment (recently installed at Futtyghur) argued to the Medical Board and the Governor-General in Council that it should be he, as a military surgeon, who be responsible for its supervision. Webster was at pains to point out that it was the military purse paying for such hospitals. He noted that had the government intended the hospitals to be civil institutions, authorisation for their funding would have come from the local magistrates.[80] This argument, and the progress of the Futtyghur lock hospital from civil to military institution, also suggests the ways in which military concerns increasingly superseded those of a civil nature. Mr Reddie, the regional surgeon overseeing both Cawnpore and Futtyghur, was soon drawn into the argument. He noted the great disparity in European troop numbers between the two stations in what appears to have been an attempt to encourage the redirection of resources to the lock hospital at Cawnpore (where he argued, the potential health risks to European men were far greater). After listening to this debate, however, the Medical Board decided that in view of the '... very limited advantages derived from it' (i.e. the relatively small number of European troops who would benefit from it), the Futtyghur lock hospital be abolished.[81] The hospital at Futtyghur is an intriguing illumination of the change in attitude which took place from the hospital's initial establishment in 1797 to the arguments in 1827 surrounding its administration. Over this short 30-year period, a more powerful military demand forcefully subverted any nascent interest in the health of Indian civilians. In a similar manner, we can see the shift to reflect more explicitly military concerns which prioritised the health of the European soldiery at stations that European troops had vacated (but where sepoy troops remained stationed) such as Palamcotta, Sholapore and Wallajabad[82] where, in 1829 it was found expedient to abolish the lock hospitals, as they were determined to be 'only requisite at stations where European troops are quartered.'[83]

Excuses, solutions and the production of racial and cultural stereotypes

Faced with a growing pressure to explain the continued failure of the system to produce the results predicted, medical officers resorted to a series of excuses. The majority of surgeons who reported on the progress of the lock hospitals strongly supported the system, despite the fact that most lock hospitals failed to significantly improve the men's health. In providing their explanations for the dramatic shortcomings of the system, medical and military men consolidated a set of ideas about the social, racial and sexual character of various groups of Indians. These excuses are therefore important to any analysis on the development of racial theory and racism across the empire.

We can divide these explanations into two broad categories. The first were relatively straightforward and stressed the more material shortcomings of the system, such as the lamentable physical state of the hospital's premises or the poor quality of the food provided. For example, one Madras surgeon suggested that greater numbers of women might be induced to agree to treatment in the lock hospital if the accommodation was made more comfortable.[84] These explanations were far fewer in number than those of the second group; these latter excuses were not only more common, but were more complex and insidious. These identified the Indians involved in the system as responsible for its failure, from the prostitute who evaded examination to the peon who neglected to bring her to the lock hospital by force. This set of excuses rested on claiming to detect a pattern of deeply entrenched moral and social corruption in Indian society. The possibility that the women's avoidance of the lock hospital was a natural, logical response to an oppressive and invasive system was never entertained. Instead, they insisted that specifically 'Indian' characteristics derailed the system. These qualities were broad-ranging and touched on many areas of life, from religious practices and caste, to a supposed tendency to both obstinacy and criminality. Surgeons assured their readers in authorial, medical tones, that such flaws were integral parts of the Indian being. Attacks, such as those levelled against alleged 'caste prejudice' and 'Hindu superstition' amalgamated with the newly emerging theories constructed about 'race' (as will be discussed in Chapter 3). The reports by surgeons and commanding officers that stressed the supposed duplicity of the peons and the 'dangerous' sexuality of certain groups of Indian women were expanded and such traits inserted into stereotyped tropes describing Indian society.

Among the most commonly cited reasons given for the failure of the lock hospital system was the fact that the predicted multitude of women voluntarily submitting themselves for examination and treatment had not materialised; only the poorest women, or those dying from starvation, entered lock hospitals voluntarily.[85] Military officials assumed that this dearth of sufficient numbers of women under treatment meant that there were untold numbers of undetected, diseased women at large who were infecting the European soldiers with venereal disease. Officials blamed this fact on both the women themselves (their 'selfish' behaviour prompted them to avoid the treatment and examinations required of them) as well as on the police and peons hired to enforce the system. This led to two further (contradictory) explanations. The first argued that the high number of women remaining 'at liberty' resulted from the fact that peons did not possess the necessary coercive power which would allow them to take in 'infected' women. Officials envisaged that a stronger force of peons would serve two primary functions: the detection (and perhaps informal examination) of women suspected of disease, and the forced escort of such women to the lock hospital for examination and treatment. The second argument held that peons were corrupt and had allowed women to evade detection in the expectation of personal pecuniary gain.

An 1814 report by the Madras Medical Board suggested that the problem of low 'voluntary' attendance resulted from religious concerns. The Board argued that inside the lock hospitals, surgeons had indiscriminately inter-mixed Muslim and Hindu women and different caste groups; this, they asserted, had deterred certain women from freely applying for treatment.[86] To remedy this, the Board suggested the internal division of all future lock hospitals, with a section each for Muslim, Hindu and 'pariah' women.[87] Here, it seems that the Board were conflating 'religious' and 'caste' concerns. Yet, beyond the segregation of the 'pariah' women into a specific ward, no further subdivision in the Hindu or Muslim wards was called for. This fact, and the selective manner in which both 'caste' and 'religion' were utilised by surgeons to achieve their own goals, suggests not just an imperfect understanding of these concerns, but a wilful superficiality. Insistence on the importance of caste considerations seems especially ironic when we consider that the European models of prison and hospital (and the amalgamation of the two in the form of the lock hospital) were alien concepts to India. Such 'caste' concerns were just as easily dismissed as unimportant, as was the case in Almorah, where, Assistant Surgeon Mitchelson of the 23rd Regiment complained, a lock hospital building had *not* been established and the women were instead treated in their own houses. Mitchelson noted that this was either due to the 'prejudices' of the women or to 'their peculiar situation in regard to cast, being chiefly Hindoos.'[88] However, in either case, he found both 'excuses' objectionable, as the women remained free and could continue to (in his view), communicate disease, to the soldiers.

In 1819, the Bengal Medical Board acknowledged that the women viewed detention in the lock hospital as being severely 'degrading'. However, instead of linking this to caste (or, for that matter, personal dignity), the Board assured government that this feeling was primarily due to the women's personal selfishness. While they acknowledged that any time spent in the lock hospital severely impacted a woman's ability to earn her livelihood, they dismissed this as unimportant.[89] Here, the language used by officials to describe protests and resistance to lock hospitals (like that used to describe protests of prison conditions or enforced vaccination), reduced the legitimate concerns of individuals to 'native prejudice', 'superstition' or 'greed'.[90]

Attacks that blamed 'native superstition' for the lock hospitals' failure closely mirrored those which targeted specific Hindu and Muslim practices. The influx of missionary groups after 1813 bolstered these offensives on religious traditions, lending their disapproving voices to the broader chorus of criticism. After a few disappointing years of the system's operation, surgeons suggested that the police's inability to interfere with Hindu 'religious institutions' (in other words, temples and the dancing girls within) was a factor in the lock hospital's failure.[91] As discussed in the previous chapter, while earlier officials had been wary of interfering with places of Hindu worship or offending the 'prejudices of the natives' for fear of incurring public unrest, this stance had weakened considerably by 1810. In 1809, outlining the causes of the failure of the lock hospitals, the Madras Military Board proclaimed

> That the effective measures have not been adopted, as used to be resorted to in cases of distress, before Lock Hospitals were established, is probably owing to the necessary caution held within the General Order... *not to interfere with the Religious Establishment of the Natives or offend their prejudices*.[92]

In this case, it was suggested that not only were temple dancing girls practising prostitution, but that other diseased women sought (and were granted) sanctuary in the local temples by 'posing' as temple dancing girls, thus evading capture. Here again, it was felt that 'superstition' and deceit prevented the success of the western 'cures' on offer. Medical and military officials dismissed the women's (quite legitimate) reservations regarding these dubious methods of treatment and instead accused them of duplicity and of hiding behind 'corrupt' religious practices in order to escape incarceration.

In the same way that European officials selectively utilised 'caste', certain 'native' practices were upheld as 'traditionally' acceptable. The Madras Medical Board, keen to provide solutions to the problems posed, suggested more strict punishment for the non-attendance of women at lock hospitals. Falling back on what they deemed to be 'traditional' Hindu methods of castigation, the Medical Board suggested that any woman who continued to practice prostitution, while knowing that she was diseased, should, upon discovery, have her head shaved and be drummed out of the cantonment. The Medical Board were confident that this measure would not provoke outrage from the Indian population, proclaiming '[we] feel assured that so far from such a measure proving offensive to the Natives *it is one which accords precisely with their own system of police* in similar situations.'[93] Supporters of the lock hospital system were nothing if not opportunists. They were content to use any evidence that would support their demands – regardless of any contradictions. Hence, while the system itself was far removed from 'traditional' Indian methods of treating patients (as well as punishing 'criminals'), these Madras surgeons selectively appropriated what they believed to be 'traditional' punishment to apply to 'errant' women.

Military and medical officers also lamented that many women sought treatment privately from native practitioners. This is perhaps even more ironic when we remember that many native practitioners were using similar, if not identical, medicines as their European counterparts.[94] While the treatment on offer from these practitioners was perhaps no less unpleasant than what was being offered at the lock hospital, no doubt the advantage the former had (in the eyes of many women) was that they were able to return to their homes after their treatment. The dismissive comments made by European officers, which reduced women's concerns and decisions to 'superstition' and 'native prejudice' were repeated *in absurdum,* as if to establish without a doubt the selfish native character which fatally disabled this 'benevolent' system.

The failure to achieve the desired number of attendees led to a further distinction in the minds of surgeons and commanding officers. Some supporters of regulation increasingly sought to portray the existence of two different groups within the overall 'prostitute' category; the first were 'respectable' prostitutes who *voluntarily* registered themselves, and the second were 'disrespectable' women, who slept with the soldiers 'covertly'. The vague descriptions that surrounded the second group were usually no more specific than the suggestion that they lived in 'haunts' and plied their trade in secret. It is possible that at least some of the women described as 'covert' prostitutes were those mistresses of soldiers or officers.

The police, peons, bazaar *kotwals* and 'matrons' who formed the surveillance and enforcement end of the system were seen as both essential and undermining at the same time. Medical and military officers frequently called for a 'well constituted and efficient *medical* police'.[95] Ironically, however, surgeons just as frequently vilified the very same police, peons and matrons as being criminally complicit in the women's intrigues, concealing diseased women and/or accepting bribes. A letter to the Bengal Military Board from the Medical Board noted that women suspected of being diseased were

> ... often assisted by the Cutwal [sic] or head Cantonment police officer, who acts consistently with the rapacity of the Native Character, and does not scruple to derive emolument even from this impure and abominable channel.[96]

Frustrated surgeons relayed suggestions of the widespread incidence of bribery and collusion in relation to the peons to the Medical Board. It was suggested that the 'higher class' of prostitutes, finding the examinations distasteful, bribed the peons to be exempted from the examinations, or alternatively, simply fled the catchment area.[97]

Mr Pears, an Assistant Surgeon at Berhampore, added his voice to the chorus of surgeons complaining about the actions of native intermediaries. In reporting his discovery of a diseased underage girl in the course of his examinations, Pears angrily attacked the matron in charge of the lock hospital, noting that the girl was '... kept from examination by the woman superintending and only produced on a threat of being reported to the Commanding Officer.'[98] Thus, the peons and matrons were seen to require close and careful observation to ensure that the deceitful elements of their personalities were held in check.

It is likely that a number of the problems connected with the operation of the system resulted from mutual cultural incomprehension in terms of the understandings of 'prostitution' and 'crime' in Europe and India. While local pre-existing structures of power remained in force and were encouraged by colonial rule,[99] as was the case of the peons and bazaar *kotwals*, the acts requiring 'policing' now changed. Indigenous conceptions of 'crime' and actionable, punishable offences were often at variance with European notions of the same. In the Mughal period, punishment for a crime was often determined by weighing a number of factors, including the offenders' 'rank' or social status with local norms and conditions.[100] In contrast, European notions of 'justice' supported fixed, immutable penalties. In much the same way, systems of policing and justice underwent radical change with the assumption of Company control.[101] The informal systems of policing that existed prior to the arrival of Europeans, simply put, were not overly concerned (if at all) with the actions of women who were later deemed to be 'prostitutes'. The women's 'conversion' to occupy the criminal category was primarily driven by the British. In this light, the frequently cited unwillingness of the *kotwal* and peons to round-up the women might be owing to the fact that they were only gradually coming to see the women as 'criminals'. The bribes which medical and army officers insisted were commonplace also suggest the on-going consequences of a system which was formed to cater to very different needs, not to those of the often unruly bunch of young European soldiers in whose name these actions were now justified. While differing understandings of the women's position may have informed the actions of the police, the continuing push by the army towards a greater integration of police and medical duties highlights their firm association of 'prostitution' with criminality; this notion would, in time, be absorbed by broader swathes of Indian society.

The lock hospital materially represented the fusion of the hospital and prison, combined to achieve the ultimate goal of military security. The system was emblematic of the dramatically changed ideas of the place and, indeed, value of 'prostitutes' in society. While, as has been discussed, these groups of women were once seen as part and parcel of the functioning framework of society, whose roles ranged from cultured courtesans to public dancers, European notions of 'prostitution' were now imposed and the women seen as performing a purely sexual role which related to a different and very narrow conception of sex as penetration. The cantonment prostitutes, serving as they did the army, were seen to merit the closest surveillance. The lock hospitals themselves physically reflected this changed view. Officers often described the buildings as 'compounds'; constructed with a high wall surrounding the grounds as well as an internal well. Like jails, each hospital was meant to be isolated and self-sufficient.[102] This vision can be seen in the construction estimate for one lock hospital in Madras in 1815. The St Thomas' Mount lock hospital was designed to hold 25 patients and included both a bath room (presumably so the women would not have to travel outside its walls to bathe) and a guard room under the same roof as the wards. Most important, however, was the 'compound wall', to prevent the women escaping, as well as determined soldiers from getting in, which would surround the building.[103]

Control and compulsion remained central to the system and lock hospital supporters constantly belaboured the necessity of both. In most stations after 1818, the use of peons was severely restricted in an effort to appease the Board of Commissioners and cut the operating costs of the lock hospitals. However, the system's supporters resented the reduction in the number of peons and this became a point of contention between officers and surgeons on one side and the Board of Commissioners on the other. This remained a sensitive issue and was central in the arguments which surrounded the proposed abolition of the system in Bengal in 1831. A letter written to the Adjutant General begged for the retention of the peons, noting that most of the women in lock hospitals were 'taken up by the Peons and ... they rarely apply for relief until by severe suffering they are induced incapable of pursuing their usual avocations.'[104] Unmoved, the Medical Board in Bengal began to press for the abolition of lock hospitals in that presidency in 1831. The failing, expensive lock hospitals could not escape the fiscal belt-tightening which categorised Bentinck's Governor-Generalship. In a period where savings were sought in myriad ways (including, controversially, the cutting of officers' *batta*[105] allowances), any plan to convince government to dedicate further funds to expand a failing system was unlikely to succeed. Nevertheless, surgeons and officers reacted with immediate resistance and instead demanded an increase in the number of peons, blaming any failings on their reduced numbers. Lieutenant Colonel Green, commanding Her Majesty's 20th Regiment in 1831 repeated this point when he wrote despairingly that

> ... since the number of Peons have been reduced it has been impossible to prevent the women coming within the cantonment in consequence of the number of Bazars and Villages in the vicinity ... entertaining the six peons as before at this station, where the disease is of a virulent nature will be found by the Government ultimately to have the most beneficial effects, in addition to keeping up the efficiency of the Regiment.[106]

In a similar manner, Colonel Sullivan, commanding at Poona[107] in 1833, observed that there were 39 women in the lock hospital, and, given the much greater number of diseased men at the station, declared that it could safely be assumed that the women in hospital '... form but a small proportion of those who prowl about these open and extended cantonments, to the destruction of the men's health.'[108] In addition to serving as a plea for further peons, Sullivan's description of such dangerous, elusive 'prowling' women highlights the additional concerns which were raised regarding the proximity of cantonments to areas of potential temptation and ill-repute. Not only was it thought imperative to physically separate such women from the rest of Indian society, but demands were now made (discussed in greater detail in Chapter 4) for increased levels of physical control in and around the cantonment, all in the name of fortifying military health.

The supposedly limited radius of control around the cantonments was a frequent complaint of military officials and surgeons alike. Surgeons blamed faltering lock hospital results on the refusal of magistrates to grant them permission to arrest women who had fled to adjacent towns. This suggests a certain tension between the civil and military arms of the Company. Mr Gibb, a Superintending Surgeon at Meerut wrote to the Medical Board that

> ... the Medical Staff in charge [of the lock hospitals] have no control over the public women of the adjacent cities, who have constant intercourse with our troops and generally harbour their Cantonment associates when diseased and desirous of eluding inspection ... Were the public women in all the great cities contiguous to our cantonments liable to inspection ... the Hospital for Females might answer the wise and benevolent intentions of Government to the full extent.[109]

In response to concerns such as this, the Bengal and Madras Medical Boards suggested that the government authorise interference in the public and judicial sphere on the grounds of 'public health'.[110] However, such intrusions into civil jurisdiction were uneven. Despite their best efforts, military and medical officers were never wholly successful in cowing civil magistrates into cooperation with their designs.

During the 1820s and 1830s, army officers and military surgeons worked in conjunction to avoid any further reductions to the lock hospital system by the government. This often included employing 'creative' ways of establishing unofficial regulatory systems. Commanding officers hastily ordered temporary and unofficial hospitals and retrospectively requested their authorisation. Reports that pressed for the expansion of the system constantly belaboured the 'urgent' and 'pressing' need for such hospitals. In an 1825 exchange between the Assistant Surgeon at Rajcote,[111] Mr McDowell and the Commanding Officer, Major Gordon, we see just how swiftly such systems were established when the interested parties were sympathetic to the system. McDowell complained of the 'encreased [sic] number of public women who at present inhabit the camp bazar at this station, and the inefficacy of such measures as the Bazar master has had it in his powers to adopt for obviating the inconveniences arising thereupon.'[112] Responding immediately to this rather vague request, Major Gordon authorised a temporary lock hospital the next day, highlighting in his letter to the Assistant Adjutant General the 'extent of the venereal disease at present prevailing amongst the soldiery and Camp followers (both European and Native)' and warning of the pernicious effects of such a 'dangerous and injurious a disorder' on the troops.[113]

At stations where lock hospitals had not yet been established, surgeons and officers who supported the system were adept at wielding its tools in order to reproduce *de facto* arrangements. They skilfully utilised the key phrases proven to provoke action and unofficially patched together the means to operate their own systems. Unofficial actions were common and varied; they would serve as the basis for similar actions that took place during the period of the lock hospitals' official abolition after 1835. These ranged from unofficial examination to 'temporary' establishments, or those paid for out of bazaar funds. At Dapolie, the Superintending Surgeon ordered the impromptu examination of all 'public women' in the bazaar areas in October 1826.[114] At some stations, we find some indications that camp followers were unofficially targeted for inspection.[115] In Almorah in 1827, where a lock hospital had been authorised, but not yet built, the surgeon conducted visits to the houses of the 'public women' with roughly the same regularity as at stations where lock hospitals were physically present.[116]

In this way, many lock hospitals and their associated supporting police continued to function without the explicit sanction of the Medical Board before 1831 (and, as will be discussed, it is likely that many continued to do so after that date). At Berhampore, Assistant Surgeon Pears blamed the recent decrease in the number of women attending the hospital on what he considered the discordant day-to-day operation of the station's medical facilities, which resulted from the neglect of the station's Superintending Surgeon, Mr Keys.[117] In the course of his argument, Pears also revealed that Major Broughton, the Commanding Officer for Berhampore, had ordered the establishment of an additional, unofficial venereal hospital located in the bazaar for 'women of pleasure' where Pears was ordered to attend weekly.[118] Over the 1820s and 1830s, as calls for economy by the Medical Boards and the Board of Commissioners grew, commanding officers increasingly relied upon such arrangements. The officers who ordered such informal systems were absolutely convinced of their utility and wished to avoid the necessity of proceeding through time-consuming official channels to await explicit authorisation. When not officially sanctioned, these impromptu venereal hospitals and regulatory systems were often paid for by cantonment funds.[119]

The increasing association of women deemed 'prostitutes' with crime was due in large part to the actions of a military-medical alliance which existed in the early nineteenth century. Not only did it determine how they were treated but it shaped the legitimate means by which the military could exert control over them. Assumptions that stressed the otherness and criminality of the women suited the needs of both groups. These distinctions ring-fenced women deemed to be a 'threat' from the rest of society and signalled that their treatment and punishment in the lock hospital was somehow justified. Moreover, they implied in patriarchal terms that the 'natural' guardians of their actions should be police and military officials and justified the women's punishment – through fines or imprisonment, when they sought to avoid the lock hospital.

For much of the period, the Board of Commissioners accepted these portrayals and demands and seemed willing to let hospital numbers increase when European military numbers merited it. They did, however, entreat the military and medical boards to seek economy in the operation of these hospitals. The ways in which such economies were interpreted by the medical boards were telling. The Madras Medical Board, appealing to this demand, suggested that only such women 'who are known to frequent the European lines and the women of soldiers who are found on examination to be infected' be admitted for treatment.[120] Unless their treatment would directly correlate to a marked increase in the health of European troops, the women were not deemed to be 'fit subjects' for the hospital and the Company was uninterested in paying for their 'cure'.

The definite link held to exist between lock hospital admissions and the preservation of the health of the European soldier was not strong enough to prevent a rapid *volte-face* by the Bengal Medical Board. While once firm supporters of the system, they were persuaded by the findings of the Inspector General of Hospitals, William Burke. Burke's 1831 report on the working of the lock hospital system suggested that it had completely failed in its aims and, had indeed prompted an *increase* in admissions among the men. Accordingly, the Board now argued for the abolition of the system.[121] Governor-General Bentinck similarly embraced Burke's report and also made clear his disapproval surrounding such methods of compulsion.

Conclusion

While the Board's rapid change of heart is at first puzzling, when situated within the broader context of the severe economic pressures and budget deficits that the Company was suffering from at the time, the reasons behind this decision become clearer. The greatest concern about the lock hospitals remained their cost, which, when measured against the relatively small reduction in disease levels among the troops was deemed too much to bear. These hospitals, the Bengal Medical Board reasoned, were simply too expensive to maintain given their inconclusive, disappointing results. The closure of the lock hospitals promised a significant annual savings for the Company at a time when fiscal concerns and budget-balancing were all important. In late June 1831, Governor-General Bentinck ordered the abolition of lock hospitals across the Bengal Presidency. However, it would not be until 1835 that Madras and Bombay reluctantly followed Bengal's lead and officially abolished the system there.

While the timing of the abolition of the lock hospitals might at first encourage the association of the action with the supposed period of liberal 'reform', their closure was not the result of any liberal or humanitarian concerns, despite a superficial gloss suggesting as much by Bentinck. Instead, like many reforms of the period, the lock hospitals' lack of economy determined their fate. However, despite the system's abolition, security and economic concerns regarding the health of the troops continued to be used by army officials seeking alternative means of preserving such health. While it would take another 20 years before military and medical officers were successful in their demands to officially re-instate the system, it never disappeared from the medical or military agenda.

The lock hospital system in India was born out of a fear that monitoring or punishing the men for contracting venereal disease would lead to mutinous behaviour. Yet the costs involved in maintaining the European soldiery in India were too high to ignore the risks posed by venereal diseases. Based on the assumption that the 'low-class' soldiers 'naturally' visited prostitutes to satisfy their sexual needs, surgeons and commanding officers instead targeted these women for control, treatment and punishment. The women reacted with insubordinate behaviour of their own – avoiding detection, inspection and treatment whenever possible. Like the 'everyday forms of peasant resistance' described by James Scott, the women subtly but consistently rebelled against the system.[122] Yet, while such behaviour might have been considered natural and understandable in the men, frustrated surgeons insisted that such behaviour (in the women) was the result of certain specific racial and cultural characteristics.

While this defence was intended to explain the poor performance of the lock hospitals in reducing levels of disease among the European troops, its logic soon spread beyond discussions of the system itself and was absorbed into a broader discourse emerging on race and morality. The 'truths' about the Indian character generated by surgeons intent on justifying lock hospital continuance were transmitted to Europe, having been severed from the very specific crucible in which they were formed.

3
Medicine and Disease in the 'Age of Reform'

In Britain, the 1830s was a period of significant legal and intellectual change – the 'Age of Reform'. The growing influence of liberalism, evangelism and utilitarianism reverberated across the political spectrum. This has led some historians to apply the same title to the period in British India. The 'reformers' of the Age of Reform have often been characterised as a collaborative band of 'Anglicists' and free traders vying against conservative 'Orientalists'.[1] The literature on the period usually examines the reforms implemented by the former group that included those centred on education, women and the law.[2] While not wishing to minimise the significance of these reforms, this chapter examines a different current running through this period. While pushing forward territorially and ensuring that their ability to collect revenue was secure, the Company continued to be concerned with the ability of its army to remain powerful enough to function effectively. As such, the Company's decision-making in the 1830s remained largely determined by its military, medical and fiscal concerns.

There were three prominent forces that moulded the period: the economic woes which threatened the Company, the prevailing Anglo-Indian notion of military-fiscalism (which meant that the army still drove most decision-making and spending), and the health risks faced by the growing legion of European troops in India.[3] When seen in the light of these concerns, a very different set of 'reforms' emerges. The nature of the Company's rule meant that while the ideas of the liberals, utilitarians and evangelicals grew in popularity in metropolitan Britain in the 1820s and 1830s, a preoccupation with commercial, security and revenue demands still dictated Company policy in India. The costs and implications of a weakened army represented a severe strain on a ruling system dependent on military force and organisation. Thus, while the 'Age of Reform' was one during which the foundations for later liberal intervention in India were laid down, the most substantive reforms implemented during the period remained fiscal and military in nature. The needs of the military, often with regard to the health of the troops, superseded all others. The impact of these reforms on Indian society, while not immediately obvious, was extremely significant.

It was in the context of this period and in response to the Company's changed role that military, civil and medical training underwent a significant change. The Company's colleges at Haileybury and Addiscombe opened in the early years of the century with the aim of providing a uniform cadre of men to send to India.[4] In addition, new policies were introduced which more firmly professionalised medicine from the late eighteenth century, and in so doing, subtly altered the character of medical men departing for India.[5] Focusing on the late eighteenth and early nineteenth centuries, what follows shows that while this move towards standardisation was the ideal, these noble ends were rarely achieved as cadets often remained stubbornly focused on less salubrious extra-curricular activities than on their studies. Nevertheless, these institutions were highly effective (albeit unintentionally) in creating a bond between their cadets.

In some instances, the modernising demands of the liberals and the Directors' desire to cut costs were successfully married. The new Indian medical colleges, informed by the same ideology that prompted the founding of Haileybury, the College at Fort William and Addiscombe, served to train up a new cadre of Indian medical officers to serve the army at a relatively low cost (certainly in comparison to their European equivalents) to the Company.[6] The opening of these medical colleges and the rise of professional societies and their associated journals corresponded with a further change in the period. Instead of continuing in the tradition of learning *from* Indians, the nineteenth century became one in which learning *about* them became more important, with obsessive categorisations and taxonomies created about much of the natural world. Over the course of the century, the European projection of the traditional, superstitious and ignorant Indian came to dominate (and increasingly shape) decision-making in many fields, including medicine and law. This chapter explores how the demands and failures of medical and army commanders contributed to the creation of this perception during this time and encouraged segregation between Europeans and Indians.

At the same time that military and civil training underwent this formalisation, European medical institutions in India saw a similar transformation.[7] This grew out of the convergence of three factors, namely increasing professionalisation in the field of medicine; the growth and activities of the three presidency medical boards; and the burgeoning number of professional (and social) scientific societies and their accompanying journals. The same *esprit de corps* that linked Haileyburians bound European surgeons in India. In the case of venereal disease, journals and medical boards functioned as conduits to produce a 'common sense' about treatment and causes. Even in the face of the persistent failure of the lock hospitals, the system remained popular among surgeons and commanding officers. In order to understand why this was the case, we must first examine how journals and circulars were used to reinforce European military and medical demands and generate this dedication to a failed system. Through these media, surgeons projected their demands more forcefully, with one voice – which by and large argued for the return of the lock hospitals. This unified opinion was the result not just of the surgeons' own experiences with venereal disease control and lock hospital operation, but emerged out of this broader network of engagement.

Medical and scientific journals and societies served an additional purpose for European medical men in India. The theories that emerged on human difference were not simply the result of a new turn in European scientific theory. Members of the numerous professional societies that emerged in the 1830s discussed new ideas that stressed biological and social difference and measured them against their own experiences in India. The societies provided a forum to debate their findings as well as those of their European peers. This chapter suggests that theories generated out of Indian experiences contributed in significant and hitherto unrecognised ways to the broader debate on human difference, morality and disease.

Surgeons and administrators in the Age of 'Reform'

The years between 1813 and 1835 saw the twilight of eighteenth-century conservatism and orientalism and the dawn of a new administration, with servants and representatives trained by more uniform, formal means. The colleges of Fort William in Calcutta and Haileybury in Hertfordshire (founded in 1802 and 1804, respectively) were crucial in moulding a new type of Company civil servant. Addiscombe in Croydon opened in 1809 to provide technical training for artillery and engineer officers. Sent out to a remote district, civil administrators were charged with filling three key roles: peace keeper, revenue collector and judge.[8] Likewise, as Company control expanded, it required greater numbers of trained men to staff its scientific corps.[9] Yet, what these colleges would prove was that the rapidly expanding need for skilled, well-educated men, could not be met by the paltry pool of candidates which Haileybury or Addiscombe produced, many of whom were awarded their places through an unhappy combination of patronage and 'cramming' schools. This was far removed from the Company's aims for the institutions, but had much to do with their other (more pressing) priority – economy. Lofty goals were one thing, cost was another. The near-constant deficits that plagued the Company in the early nineteenth century encouraged a determined quest for cost-saving measures. In establishing Addiscombe, the Company sought to avoid the expense of educating cadets at the Military Academy at Woolwich, which was, on average £250 per cadet.[10] It was hoped that 'in-house' training would not only be less expensive but better suited to the specific needs of the Company. In promoting a standardised set of values, these colleges aimed to instil a new set of beliefs, and it was hoped, morals, into the men trained there.[11] The establishment of the three colleges was intended as a clear signal that the days of the slightly (or even explicitly) eccentric administrators, who had garnered their knowledge of India in a more organic manner (from sources which ranged from *munshis* and mistresses to Mughals) were numbered.

On its founding, the College at Fort William offered instruction in Arabic, Persian, Sanskrit, Hindustani, Bengali, Telugu, Marathi, Tamil and Kannada. Additional course offerings included Indian and English law. Similarly, Haileybury students engaged in a range of subjects, including English law, politics, finance and commerce. It was felt that these were essential for future employees.[12] It was expected that students, prior to arriving at Haileybury, would have a firm background in both the classics and the Bible.[13] 'Oriental languages' were introduced to students, with a special focus on Sanskrit and Persian. However, apart from these language classes, students learned almost nothing about India or what they might expect from their future lives there. Even Sanskrit lost some of its importance when it was made optional in 1814. After this time, only those bound for Madras were required to learn the language, while those headed for Bombay and Bengal were merely encouraged to do so.[14] In his memoirs, John 'Cross' Beames remembered that at Haileybury, 'India was not talked of or thought of except by the few who really worked, nor did we as a rule care or know or seek to know anything about it.'[15] The

Directors intended this knowledge to be absorbed during the students' leisure time. The report that ordered Haileybury's founding directed that

> The private reading of the Pupils should be partly directed to a proper selection of Books on Asiatic subjects, particularly Histories, and accounts of the Character, Manners and Literature of the Eastern People. Part of the Vacations might be usefully employed in Viewing the great Public Works in England[:] Docks, Arsenals, Manufactories and the like, of which they will know nothing but by report, before a late period in life.[16]

Given that these were still relatively young boys, their interests predictably remained firmly planted in activities other than visiting dockyards. Perhaps the most zealous among them followed the Commissioners' wishes, but the vast majority did not, preferring instead activities such as rowing and drinking for their many free hours outside the classroom. Haileybury was a place of privilege, and, as Beames' account makes clear, as long as the men did not commit any massive offence, teachers turned a blind eye to breaches in discipline and learning. Beames wryly insisted that '... while the facilities for *not* learning were considerable, those for learning, were, in practice, somewhat scanty.'[17]

Addiscombe, the Company's military seminary, was aimed at the education of those bound for the officer class. Addiscombe fed its cadets on a diet of study and drill (along with obligatory daily attendances at the chapel). Addiscombe students spent their days engaged in courses in mathematics, fortification, 'Hindustani', military drawing, surveying and civil (landscape) drawing.[18] Of these, the study of vernacular languages (seen by Addiscombe's founders as so critical) was negligible at best. Like Haileybury, Addiscombe attracted a broad range of critics. Some argued that the subjects taught were insufficient or inappropriate, while others attacked the intensive, long hours of study, preferring instead that cadets be encouraged to spend more time outdoors, engaged in 'masculine' activities such as sport. The Seminary's many inadequacies meant that it never achieved the standards initially set for its recruits and in time, had to lower the graduation requirements.[19] Like Haileyburians, Addiscombe's cadets were criticised for their lack of discipline and bad behaviour. One anonymous critic argued that the practice of

> Keeping the cadet a prisoner during the sunny hours of the day, and emancipat[ing] him at night-fall [was] responsible for a sad catalogue of grievous sins committed by the students, which, but for the

> vicious regulations in force at the institution, would not and could not be committed.[20]

What these 'vicious regulations' amounted to were the long (indoor) study hours. Lamentably (in the eyes of the above critic), the Addiscombe day started and ended with mathematics. In fact, mathematics consumed 22 of the 54 hours in the academic week.[21] He argued that cadets naturally rebelled against these long hours of study; in this light, he reasoned that the boys' bad behaviour and turn to drink and prostitutes, was a 'natural' response to an overly rigid study routine. To highlight his point, the critic insisted that the survey class, when sent on occasional field trips to '... survey the Norwood Hill or the Brighton Road, were generally better disposed to take the measurement of the parlour of the *Rose and Crown* or the *Jolly Sailor.*'[22] However, the Company's Directors did not seem overly concerned with expending extra resources to crack down on this lack of discipline, improve the curriculum, or attempt to reform the morals of the students of either Addiscombe or Haileybury (much less the average soldier), nor was it in their interests to do so. What both colleges did, however unintentionally, was to create bonds between their students. Whether studying military drawing or forging invitation letters from fictitious London relatives for each other,[23] the boys formed associations at colleges that would in many cases, last their whole careers. This was certainly true of Beames, who met his lifelong friend and correspondent, Frederick Eden Elliot while wandering through Haileybury's quad. Beames proudly remembered that Elliot '... broke all the college rules whenever it suited him.'[24] Like Beames, Elliot went on to serve in various civil postings in India.

In contrast to those would-be Civilians or officers, medical men wishing to serve in India had no one designated school to attend. In the Company's early years, many of the men entered into service as surgeons on Indiamen, the merchant ships operated by the Company which travelled the route to and from India. The Company's Directors awarded medical appointments in a similar manner to civil positions, that is, through patronage. It was expected that surgeons or assistants would be examined in England before their departure. Medical applicants were *expected* to possess a diploma indicating their readiness for service. However, the Company (or Crown) did not formally regulate this practice until the early years of the nineteenth century. The position of Surgeon's Mate in India was attractive enough to warrant substantial competition for a diploma course offered at the London College of Surgeons between 1745 and 1800.[25] From the position of Surgeon's

Mate, a determined and dedicated apprentice could expect to progress through the newly formalised medical structure in India towards increased rewards.

By the early years of the nineteenth century, medicine in India had become a more attractive occupation for ambitious young men, in part due to the formal structures imposed on it in England, but also those changes occurring on the ground in India. In 1764 the Bengal Medical Service was established and the other presidencies quickly followed suit. The introduction of a formal medical service ensured a more regular establishment with fixed grades and rules for promotion.[26] In 1804, King George III issued a warrant raising the pay of medical officers in the army in the hopes of encouraging more 'able and well educated persons' into its service.[27] This and the founding of the medical services corresponded with a large increase in the number of European surgeons in India. In 1785, the official medical establishment for the whole of the Indian Army amounted to a mere 234 men. By 1825, this number would grow to 680.[28] By 1822 the Company stipulated that all applicants for the position of Assistant Surgeons were required to possess a diploma *either* from Edinburgh, Glasgow, London or Dublin.[29] However, it was not until 1863 that the dedicated army medical school at Netley was founded.[30]

The three medical boards, situated in each of the Presidency capitals, regulated both the men in the field and general medical direction in India, meeting weekly to discuss best practice, current medical thinking, and the condition of the troops (including any specific outbreaks, such as syphilis or cholera). Through their constant communication with surgeons in the field, these Boards were instrumental in promoting new medical theories and practices based on specifically Indian concerns. Those surgeons who had distinguished themselves, or had spent long years in service, staffed the Boards. These higher medical positions were relatively lucrative. The annual salary of a surgeon serving in the infantry was not substantial; in 1831 a Bengal surgeon could expect to earn Rs 4,005 a year.[31] However, the reward for men who, through long years of service, reached the rank of Superintending Surgeon was an annual salary of Rs 15,600.[32] As a result, there were considerable incentives for ambitious young men to consider a career in medicine in India.

The increased demand for officers, administrators and medical men in India was in no small part due to the rapid territorial expansion of the Company in the late eighteenth and early nineteenth centuries. The East India Company had, through its aggressive policies, gained direct or indirect control of much of India by the early years of the nineteenth century. While territorial expansion certainly meant a larger revenue collection area for the Company, it also necessitated the costly growth of civil and military staff. This, in turn, led to an increase in medical staff to minister to their needs. These campaigns came at a high cost. Following the second Anglo-Maratha war of 1803 to 1806, the Company's debt doubled.[33] Rising expenditures forced the Company to appeal for monetary assistance from the British government. In the years from 1814 to 1823, the remittances sent back to England to pay for the Company's 'Home Charges' did not cover its operational expenses. At the same time, the Company saw its commercial profits plummet.[34] The rapid growth and bloated state of the Indian civil establishment between 1813 and 1825 placed Company revenues under further strain.[35] The First Anglo-Burmese war from 1823 to 1826 dented Company coffers to such an extent that the Indian Government teetered on the verge of bankruptcy.[36] During these three years, military charges jumped from approximately £9,117,500 in 1822–1823 to £12,824,100 in 1825–1826, with the cost of the war estimated at £4,800,000.[37] These fiscal pressures loomed over the Company's decision-making throughout the 1820s and 1830s. Such financial strains made it imperative that the army minimise its operating costs wherever possible and this pressure was certainly felt by these new legions of medical men. One central strain, which the Company desperately sought to avoid, was the cost of soldiers who were 'lost' to the army through illness or 'invaliding'. Invaliding implied long-term sick-leave or, indeed, the men's permanent discharge (and return to England at the expense of the Company) as a result of their depleted health.

European surgeons and military officials formed an alliance to respond to what they perceived to be the potential medical hazards which surrounded the troops. Their responses illuminate the scant regard paid to many of the notions propounded by the Age of Reform. They suggested instead a different set of reforms, based on the needs of the military. No amount of missionary or liberal pressure made much difference to decisions involving security or tactics. Indeed, in these there is little evidence to suggest that reformers' demands had any resonance at all, except, perhaps in altering the language officers used. Instead, like much else in the 'Age of Reform', a clear cost-benefit analysis dictated most decisions.

The Company's deficits were compounded by the economic depression of the period. The 1820s saw economic stagnation and liquidity problems, which were further complicated by the failure of the indigo crop in 1826, and the corresponding collapse of a number of

Agency Houses in 1827–1828.[38] In the five years prior to Lord William Bentinck's appointment as Governor-General, the Company's annual deficit had averaged £3 million.[39] As a result, when Bentinck returned to India in 1828, it was with explicit instructions from the Board of Directors to curb expenditure in both the army and administration. The Company desperately needed to reduce its charges. Without such a reduction, not only did profitability decrease, but there would be no surplus for either expansion or for other proposed reforms such as education or infrastructure. As a dedicated liberal and utilitarian, Bentinck hoped to utilise his position to enact a number of political, social and legal reforms. However, the most far-reaching of the changes he implemented were not those directed at the Indian populace, but instead at the Company's finances and army. These involved a series of cost-cutting and budget-balancing measures. Even the seemingly less controversial political reforms which Bentinck approved, such as those directed at infrastructure and public works, were often blocked by virtue of the depressed state of the economy. The demands of free-traders and liberals were simply silenced by the economic situation.[40]

When it came to lock hospitals, Bentinck found an opportunity to combine his Benthamite reforming zeal while cutting costs. After William Burke, the former Inspector of Hospitals in Calcutta, released his report on the lock hospitals which suggested that they had failed to achieve their goals, Bentinck ordered their abolition across Bengal. Bentinck was horrified by the compulsion used to restrain women in the lock hospitals. Moreover, he took the view that the moral improvement of the soldier (through offering him different diversions) was a much more desirable means of fighting venereal disease.[41] However, perhaps most critically in this fraught economic climate, the closure of the lock hospitals promised an annual savings to the Company of Rs 30,000.[42]

The Company was also able to combine its drive for economy with a desire to promote a 'modernising' image in its plans for a new medical college at Calcutta. The Native Medical Institution, established in 1822, trained Indian and Eurasian men for posts within the medical services; here, students were taught elements of both Indian and western systems in the vernacular.[43] The creation of the Native Medical Institution was, of course, driven by specific, military-led concerns. Many of its early students were the sons of sepoys and on graduation, students were required to serve in the medical corps for 15 years.[44] The East India Company also supported the study of unani and ayurvedic medicines through its patronage of such institutions as the Calcutta Sanskrit College and Calcutta Madrassa (where each was taught alongside translated European texts). However, by the 1830s this support had begun to ebb.

In 1833, in line with the Anglicist-leanings of Bentinck (and his advisors), a committee he appointed to examine the subject suggested that an establishment dedicated exclusively to European medicine and taught in English replace such studies.[45] In addition, it recommended the creation of scholarships for Indian students to study medicine in England for five years; upon their return they would take up teaching posts at the new college.[46] In 1835, the same year in which Thomas Macaulay produced his 'Minute on Education' proposing that English replace Sanskrit and Arabic as the language of education in India, the committee's recommendation was realised with the establishment of the Medical College at Calcutta.[47] This new college aimed to train Indian men (often from middle-class families) in European medicine; many of whom went on to serve the Company, while others entered into private practice. Teachers delivered lectures on a range of subjects, including anatomy, physiology, general chemistry, the *materia medica* and surgery.[48] The College organised observations, in the form of periodic attendance at the various Calcutta hospitals, as and when certain conditions and symptoms arose. The establishment of the Calcutta and Madras Medical Colleges in 1835 and the Grant Medical College in Bombay in 1845, implied that henceforth, European scientific and medical learning was to hold a privileged position for all Indian men hopeful of obtaining a place within the Company's medical hierarchy.

In the case of these medical colleges, the liberal ideas of the period were successful precisely because of the economic benefits that they promised. Indian medical men[49] (like Indian soldiers) required far less expenditure than their European counterparts. As the structure of colonial medicine solidified in the nineteenth century, so too did the insistence that these were members of a 'subordinate' service. They retained their heavily prefixed, hyphenated titles well into the twentieth century, many decades after Calcutta University started offering the MD degree in 1857. However, with this newly trained cadre, the Company was able to gradually expand much-needed medical support across its territories at a greatly reduced cost. As Projit Mukherji points out in his examination of the shared social identity of *daktars* (those Indians trained in western medical practices), sub-assistant surgeons, with a decade of experience still earned only Rs 150 per month following the change in service rules in 1849.[50] These men served as assistant and sub-assistant surgeons in the Native Hospitals and Dispensaries, as well as ministering to the needs of sepoys in the field and cantonments.

The *daktars* were encouraged to reproduce and reinforce European medical theories in their own practices. However, while the Colleges aimed at exporting a certain kind of European medicine in their training of Indians, in practice, medicine on the ground in India was far more complex and nuanced. Not only did European surgeons continue to borrow from the ideas and practices of their Indian contemporaries, but Indian students of the European medical colleges did not always fully adopt European practices. In areas outside of the presidency capitals, ayurveda and unani retained their popularity and colonial administrators in the *mofussil* continued to employ Indian practitioners.[51] Nevertheless, from phrenology to mesmerism, these newly educated graduates imbibed European medical theories on a much larger scale than was ever possible before.[52] Despite their emphatic support for European medicine and the great savings to the Company which they represented, these Indian medical men, like their military counterparts, were held firmly in subordination to their European peers and were prevented from serving in the higher positions available within the medical service until the twentieth century.

Essays, societies and journals

The huge volume of data collected across the presidencies and transmitted back to the medical boards provided surgeons with vast quantities of material that influenced their thinking on subjects that ranged from 'tropical' diseases to race. Competing with the flood of medical theories emerging from Europe at the time, scientific and medical observers in India sought to prove that their findings represented a distinct contribution to the advancement of knowledge. Surgeons in India, with increased access to European medical information, were keenly aware of the scientific and medical debates engaging their peers in Europe and were eager to participate. Pamela Gilbert has recently argued a similar point with relation to the mapping of cholera epidemics, noting that European medical practitioners in India used such maps to 'legitimate themselves as professionals'.[53] In much the same manner, medical surgeons, now commanding a higher degree of respect from commanding officers, attempted to emphasise their unique knowledge based on the privileged access they had to the bodies of soldiers ('foreign' and European) and, in the case of venereal diseases, to Indian women. Their theories often stressed elements of 'difference' – whether moral, cultural or physical – between Europeans and Indians. Through their writings, surgeons implicitly suggested that medical knowledge and new 'cures' could contribute to the stability of the Company's position in India as well as advance the cause of scientific knowledge.

Before examining the findings of these observers in India, it is useful to briefly consider the discourse that was developing in Europe regarding human difference.[54] In 1776, the German naturalist Johann Freidrich Blumenbach published *On the Natural Variety of Man* in Latin.[55] Blumenbach focused on theories of climate and degeneration, stressing the idea that climate was responsible for variations which existed within humans.[56] Debates on human variation in the early part of the nineteenth century centred around the idea of monogenesis, which argued that there was one centre, or 'family' responsible for the creation for the whole of humankind. This family was often literally taken to be the Christian Biblical Adam and Eve, with a concomitant argument that centred around the sons of Noah. A dedicated monogenist, Blumenbach nevertheless divided human varieties into three groups (later editions saw this number increased to five). These were: Caucasian (which he describes as the 'middle' variety); Mongolian and Ethiopian (deemed the 'extreme' varieties); and finally the 'intermediate' varieties of American and Malay.[57]

James Cowles Prichard, a scholar of Blumenbach, was also a firm adherent of monogenesis. He later founded (and became the President of) the Ethnological Society of London in 1843. However, from the point at which he completed his medical dissertation at Edinburgh in 1808,[58] Prichard distinguished himself from Blumenbach in rejecting ideas of climatic determinacy. Instead, Prichard posited, human variety arose out of the process of 'civilisation'. Prichard held up the caste system as 'proof' of how human varieties were formed. He argued that despite existing under much the same climate, by virtue of reduced intermarriage in certain groups, Indian peoples were divided into a number of 'classes', each of which retained distinct cultural practices.

Prichard, like Blumenbach, despite adhering to monogenesis, nevertheless applied a hierarchy based on colour that he implied was linked to morals and refinement. His suggestion that 'moral causes' had a bearing on human diversity is one which would appear in the theories of monogenesists and polygenesists alike. By the 1840s, polygenesists contested the theory of monogenesis, arguing instead that there were multiple 'creation' centres. However, both groups (in the years before Charles Darwin published *On the Origin of Species* in 1859) argued from the premise that species remained stable, or unchanging. The concept of evolution had been debated in the 50 years prior to the publication of *On the Origin of Species,* most notably by Jean-Baptist Lamarck in his series of lectures on the subject in 1800. However, the idea of evolution at this time was usually applied to varieties, rather than species. As such, the desire to understand how and why humans came to exhibit such variation was at the forefront of these debates. Until the acceptance of the idea of plate tectonics and continental drift in the twentieth century, scientists struggled to understand how such massive migrations of plants and animals across the seas and continents could have occurred. Darwin's experiments with seeds and plants, in which he immersed them in seawater to test their ability to germinate later, were an attempt to understand how similar plants could appear in such distant places as Africa and South America.[59]

The 'medical observation'-type essay provided European surgeons in India with an outlet to extrapolate their 'uniquely' Indian observations on moral, cultural and physical 'difference'. David Arnold has suggested that the growth of such guides on the diseases and disorders of India reinforced the colonial presence by emphasising to their readers that provided the correct (i.e. European) precautions were taken, India could be a perfectly healthy place in which to live.[60] Essays in the early years of the nineteenth century, more predictably perhaps, reflected an 'orientalist' leaning, while those produced later were much more critical of Indian society and practices. James Johnson's[61] 1813 essay, *The Influence of Tropical Climates, More Especially the Climate of India, on European Constitutions,* straddles both positions. In it, certain aspects of Hindu medical and social practices are heavily drawn upon while others are targeted for outright ridicule. Mark Harrison has argued that Johnson's essay marks the start of a movement away from the theories of acclimatisation embraced by earlier generations of medical observers.[62] Through his observations, Johnson attempted to position himself as an independent and trustworthy source on such matters.

Perhaps more tellingly for the study of venereal disease is the fact that Johnson was eager to portray himself as a moral observer, arguing strongly against Benjamin Moseley's earlier statement that, 'There is in the inhabitants of hot climates ... *a promptitude and bias to pleasure,* and an alienation from serious thought and deep reflection.'[63] This suggests a more direct criticism of the earlier generation of men, included among which were the 'orientalists' and those men who had become 'Indianised' – wearing local clothing and taking Indian wives and mistresses. Haileybury, Fort William College and Addiscombe aimed to produce a standard 'ideal' of formalised training and this older way of life was now looked upon with increasing contempt. While not countering Mosely's suggestion that such a 'bias to pleasure' existed,

Johnson refuted the climatic approach with suggestions for government and army alike. He asserted that any moral and mental laxity was due entirely to the 'monotony of life' experienced by Europeans in hot climates.[64] The daily routine of army men would be the target of frequent criticism by medical observers. It was felt that the brief, early morning drill, followed by a heavy breakfast and very few other structured daily tasks, led to boredom which more menacingly, led to mischief. Such a routine, Johnson warned,

> ... too often lead to vicious and immoral connexions with Native females, which speedily sap the foundation of principles imbibed in early youth, and involve a train of consequences, not seldom embarrassing, if not embittering every subsequent period of life![65]

In Johnson's reasoning, monotony led to 'immoral' and 'immodest' behaviour that in turn produced vice-driven threats such as venereal disease. The construct that venereal diseases resulted from 'impure' intercourse (i.e. with a prostitute) was reinforced. Johnson's essay was a plea to the Company to alter the conditions that led to such monotony.

Descriptions of venereal disease at this time were often grotesque and extreme and, more often than not, sprinkled with a healthy number of religious or 'moral' references. Such descriptions filled the pages of books and pamphlets produced in both Britain and India that promised cures or preventive measures. These pamphlets almost always situated the blame for disease on the bodies of 'unnatural' women. In his 1816 tract, *The Medical Guardian of Youth: or, a Popular Treatise on the Prevention and Cure of the Venereal Disease, so that Every Patient May Act for Himself*, Robert Thornton wrote of his belief in the necessity of warning youth of the high risks involved when they associated themselves with prostitutes,

> Youth should especially be told, that these women are almost certain of containing about them the venereal virus. These indeed are scorpions of the worst kind. What person would be so insensate, or mad, as to play with the adder, or snake, and forgetful of its power of injury.[66]

Thornton continues on with this histrionic language, assuring his readers that he had often witnessed the fast-acting ravages of untreated venereal disease destroy 'whole powers of manhood' in a few short hours.[67] Venereal prevention literature is peppered with descriptions of collapsed noses and pelvises in an (apparently unsuccessful) attempt to terrify young men away from 'unnatural' sexual activity.

In 1819, Mr Ledman, an Assistant Surgeon attached to the Sermoor[68] Battalion, concurred with Johnson's moral assessment when he wrote to the Bengal Medical Board with his concerns regarding the prevalence of venereal diseases in the area around his battalion. The 'General Dissolution of Morals', he proclaimed, had provoked high numbers of venereal infection. Ledman warned that unless quickly checked and treated, those 'unhappy victims of unnatural passions,' would ultimately be destroyed.[69] In Ledman's estimation, any plan to combat such diseases which ignored what he saw as the 'principal propagators of the Evil' – the women with whom the men associated, would surely fail.[70] If the men were fortunate enough to escape the clutches of venereal disease, any offspring they might produce were condemned in a number of other ways that usually suggested some sort of moral or racial 'degeneration'. 'Connections' with Indian women were monitored more closely and increasingly frowned upon.[71]

Medical practitioners and civilian observers in India were able to discuss their practices as well as new theories in any one of the newly formed societies which sprung up across the three presidencies in the late eighteenth and early nineteenth centuries. The Asiatic Society of Bengal, founded by William Jones in 1784, met regularly to hear original works on subjects of historical, scientific and artistic importance which specifically related to Asia.[72] From the outset, Jones instilled a rigour to the Society meant to reinforce the seriousness of the enterprise. Meetings were held weekly and only original contributions would be discussed.[73] Similarly, a number of 'Medical and Physical' societies formed in Bombay, Calcutta and Madras to discuss members' own findings as well as debate the works of their contemporaries in Europe and the other Presidencies. The earliest of these, the Calcutta Medical and Physical Society, was launched under the protective wing of the Asiatic Society in March 1823.

Medical officers stationed in more remote parts of Company territory were able to follow the debates of their peers and contribute through the printed journals that the societies produced. These societies were a crucial element in the formalisation of European medicine and science across India. To borrow Benedict Anderson's famous theory, these societies and publications were central to how the European scientific community 'imagined' itself in India.[74] Of course, this did not produce (as in Anderson's analysis) a national consciousness, but it bolstered the confidence of European practitioners across India and made them a more cohesive unit. They became bolder in expressing ideas and theories and increasingly took to professing the superiority of European medical practices over those of their indigenous counterparts. Mrinalini Sinha has argued that social clubs created a 'colonial public sphere' (which led to inclusive/exclusive notions of 'clubbability').[75] In much the same way, these scientific societies performed an important intellectual and social function, providing an arena for the men to meet and discuss ideas. These societies, and the journals they produced, were central to the formation of a coherent response to the on-going threat of venereal disease. This network was pivotal in keeping the system at the forefront of medical thinking in India even (or more precisely, *especially*) after the failure and abolition of the system in the 1830s.

Reinforcing the demand for men of science in Europe to listen to the findings of their peers in India, some of the journals moved towards a presentation of their medical knowledge as idiosyncratic of the 'East'. The introduction to the 1825 *Transactions of the Medical and Physical Society of Calcutta* asserted confidently that European men of science, '... will gladly welcome the light that may be thrown upon the past or present existence of Oriental medicine.'[76] In the Empire's eastern reaches, the field of 'oriental' or 'tropical' medicine was suggested as a distinct subject of study. European surgeons in India argued that only men like themselves, who possessed authentic, first-hand experience of such places, could truly understand these fields.

Journals such as *Gleanings in Science* (founded in 1829), printed current scientific debates as well as informing readers of the goings-on of the various societies.[77] The 1829 volume of the journal provides an excellent illustration of the subjects typically under discussion. It reports, for example, that in February at the Medical and Physical Society of Calcutta, members debated Linnaean classification, while at the Asiatic Society that month, Cuvier's *Regne Animal* was scrutinized, with supplementary findings on Indian zoology to support Cuvier's ordering provided by Major CH Smith.[78] In June, James Annesley read his work on Indian diseases[79] to an audience at the Medical and Physical Society, while in July the surgeon-scholar John Tytler presented two 'Hindu skulls' to the group.[80]

Men in Europe, like Prichard, who alluded to India in their writings, received much of their information regarding so-called Hindu 'tribes' and 'classes' from observers on the ground in India. Prichard had access to journals such as *Asiatic Researches* (the forerunner of the *Journal of the Asiatic Society*), which was sent regularly to sister societies in London, Edinburgh and Paris.[81] In addition, European medical observers in India often maintained a correspondence with their peers on the continent, sending them not only written findings, but samples which included everything from human skulls to unusual plants and religious artefacts. Through correspondence, journals and circulars, the empire (to use Andrew Thompson's phrase) 'struck back', imploring their professional peers in Britain to listen to their findings.[82]

Medical writing during the 1820s and 1830s increasingly sought to compare and contrast the bodies of Europeans and Indians. By this time, many medical writers in India confidently stated that Indians and Europeans were inherently biologically different. As 'proof' of this, they pointed to varying rates of susceptibility to certain diseases.[83] Surgeons suggested the alteration of treatments based on whether patients were Indian or European, due to their 'different' constitutions. They frequently argued that Indians were too feeble to be able to tolerate the severe effects of remedies such as purgatives, bloodletting, and, to some extent, mercurials.[84] In contrast, the 'sturdier frame' of the European could withstand such harsh methods of treatment. With increasing frequency, reports portrayed Indians as 'weak' or 'enervated'. These 'constitutional' differences contributed to the construction of theories that stressed biological and racial difference.

As awareness of physical differences between peoples spread, the notion that this corresponded to biological and mental difference grew apace. The theories of Georges Cuvier, Pieter Camper and Franz Joseph Gall,[85] whose ideas of physiognomy and phrenology stressed a direct correlation between mental ability and the shape and contours of one's head, took hold in both Europe and India. Indeed, as the popularity of phrenology waned across Europe during the 1830s, it attracted new supporters in India. The *Indian Journal of Medical Science* reprinted an article from the *Lancet* on the subject in 1834. The article relayed the details of a murder case in Paris in which a phrenologist, Mr M. Dumoutien, Professor of Phrenology to the Society of Paris, was called in to identify a murdered woman's skeleton recently unearthed in her back garden. The author recounted that Dumoutien carefully examined the skull and pronounced it had been interred for a number of years and belonged to a woman 'advanced in age'.[86] He went on to excitedly exclaim

> ... the surprise of the spectators was at its height when the Physician, continuing his remarks, commenced to speak of the character of the person whose skull he held in his hand, declaring that she was miserly, and of a violent temper, and adding other details, all of which were perfectly comfortable with the known temper and character of the missing woman.[87]

A new generation of Indian-born phrenology-enthusiasts sprung up in the 1830s, a phenomenon most likely linked to the teachings which took place in the new Indian medical colleges. *The Pamphleteer*, a phrenological journal published by the Calcutta Phrenological Society, stated its goals as being nothing less than the 'physical well-being' and 'moral and intellectual improvement of society.'[88] Produced by Indian intellectuals, *The Pamphleteer* again reinforced the idea of a hierarchy of human types. Yet, phrenology in India, like other European medical practices on the subcontinent, was never a straightforward duplication of European dictates. Instead, Indian phrenologists selectively incorporated their own beliefs and adapted European theories. This can be seen in an article entitled 'Phrenological Development of the Bengalees,' in which the author frankly asserted that 'The prevalent vice, among the illiterate, of telling fibs, and concocting false stories is also intimately connected with their cerebral organisation.'[89] However, redeeming Bengalis somewhat, he went on to state, that '... another remarkable trait in the character of our countrymen is their perseverance, industry, and artistic skill.'[90] Embracing the notion that it was possible to judge human ability by the contours of the skull, the author concluded confidently by stating that, after a close examination of the Bengali head, there could be little doubt that Bengalis were among the most intelligent people on earth.[91] Statements such as these provided a fitting complement to the phrenological ranking conducted by Europeans (who, predictably, were equally sanguine about their own position at the pinnacle of the global intellectual pyramid).

An 1839 list of questions intended for Europeans observing 'native' peoples to follow in constructing their reports further reinforced this stress on physical, moral and cultural differences. The British Association for the Advancement of Science drew up this list. (See Appendix 1 for an extract from the list.) Although not specifically a phrenological society, the Association's questions included a number that suggested the importance of bodily measurements and relative proportions, cranial and otherwise. In 1839, the editors of the *Journal of the Asiatic Society* reprinted these queries for their own members, encouraging their use for observations.[92]

The Asiatic Society's members closely followed many of these questions as they went forth to gather data. Brian Houghton Hodgson wrote a series of reports on Indian ethnology and the 'varieties' and 'types' of peoples for the *Journal of the Asiatic Society* in the late 1840s. Hodgson was a prolific collector and observer of Buddhism as well as of Tibetan and Nepalese flora and fauna. As Assistant to the Resident in Nepal (a post he was initially appointed to in 1824), he set about gathering as much information as he could on Buddhism. While living in Nepal for most of his professional life, Hodgson kept in touch with the Asiatic Society in Calcutta, sending over 125 of his papers and findings.[93] In his 1848 report on the 'Tibetan Type of Mankind,' Hodgson carefully recorded a list of measurements from a 30-year old resident of Lhasa which reflected the guidelines set out by the Committee.[94] In a later examination of the language of two 'Tamil' peoples, the Bodo and Dhimal, Hodgson found both to be 'godless'. Such must have been the case, he decided, as he could find no words in their language for such concepts as 'God', 'soul', 'heaven', 'hell' or 'sin'.[95] Yet, despite this 'godlessness', Hodgson proclaimed that both groups 'use' their wives and daughters well.[96] Overall, he found the men intelligent, honest and hard-working. The duality of Hodgson's judgement was not unique; it was replicated by many of his peers in medical writings until (at least) the 1870s. The stress on morality and 'civilisation' once again remained a central factor determining physical difference.

Such moralising tones grew increasingly common in medical observation essays. James Johnson's 1813 work enjoyed a number of reprints, as well as a revised version co-authored in 1841 with James Ranald Martin, Presidency Surgeon of Bengal (and later President of the East India Company's Medical Board), who was an ardent admirer of both Mill and Bentinck.[97] However, the 1841 edition placed greater stress on the importance of upright morals in maintaining men's health in tropical climates. In light of Martin's liberal leanings, it is perhaps predictable that this revised essay was markedly more critical of Indian society. Martin's conclusion that the vegetarian diet prescribed by Hindu 'law' made Indian men 'more feeble' and 'easy of domination,' also gained popularity in later years.[98] Earlier doctrines had similarly suggested a link between vegetarianism and temper, and it was a frequent feature in the climatic theories which professed that the natural tendency in a hot climate was for cruel and vengeful peoples. Observers proposed that the vegetarian diet of much of India had countered these malicious effects to create a certain mildness of temperament.[99] James Johnson connected vegetarianism, Hinduism and the supposed Hindu temperament, noting, 'They [the ancient Hindu "lawmakers"] probably thought, and in my opinion with good reason, that the injunction would tend to diffuse a more humane disposition among the people, by strongly reprobating the effusion of blood.'[100] By this reasoning, the consumption of animal flesh not only heated up temperaments, but led to dramatic differences in character. However, this very argument demonstrates the complexity of European medical practice in India at the time. While on the surface, it dismisses Hindu men as 'feeble' and 'submissive', the notion that diet dictated bodily 'humours' was taken directly from Hindu (and Muslim) medical theories.[101] According to humoural notions, balance was essential on every level, and the careful, managed consumption of certain foods was directly linked to broader social harmony.[102] Thus, while European surgeons had a tendency to highlight differences in the Indian 'nature', they continued to selectively borrow from indigenous medical theories.

The concept that India was a land populated by weak men (who, nevertheless ruthlessly dominated their women), who were therefore 'destined' to be subordinated, appealed to those seeking to justify Britain's growing empire in South Asia.[103] In these essays, we can clearly identify the seeds of ideas such as Britain's 'moral' obligation to rule and 'civilising' impulses. James Martin felt at liberty to depart from his medical observations in order to express his belief that with 'the all-powerful benefits of education, and the example of their European governors ... [Indians will be able to] conquer the influences of climate, and of the depraving religious and political habits of ages.'[104] Such fusing of moral, political and scientific theories became so common that by the late 1840s, medical essays on subjects such as syphilis or the dreaded 'fevers' of India, which did not contain some form of moral judgement, were rare.

The complex and often contradictory relations between Indian groups such as the Bengali *bhadralok* and European medical and military observers suggest a further underlying dynamic at work. On the one hand, the *bhadralok* were quintessential collaborators with European rule. Educated Indian men filled a significant proportion of the administrative and medical roles so necessary to the expanding Company. Many of the Bengali elite had adopted a number of the social and religious practices proposed by earlier European observers for their 'betterment' – through conversions to Christianity and a willingness to embrace the English language and European-style education. Yet, some European observations suggest an undercurrent of fear and resentment directed at their Indian collaborators. This can be seen in an article written for the *Calcutta Medical and Physical Society Quarterly Journal* entitled 'On the Medical Topography of Calcutta' which is worth quoting at length:

> The Bengallee [sic], unlike the Hindu of the north, is utterly devoid of pride, national or individual. His moral character is matter of history ... Let those who like it, follow the Bengallee in his practice of falsehood

> and perjury – his insensibility to the feelings of others – his 'perfection in timidity' – his cruelty and ferocity – his litigiousness – his physical uncleanliness and obscene worship; for my part, I prefer turning my recollections, although it may be a digression, towards that class of Hindus whose sense of military honour forms so powerful an incentive to good conduct in civil life – the up-country sepoys – a class of men at all times respectable, when justly treated.[105]

The anonymous author is at pains to assert his own superiority through his use of countless insults and backhanded compliments. His explicitly stated belief in the superiority of the European character is clear. This, along with remarks such as those linking Hindu religious beliefs and practices (including vegetarianism) to a weakness of mind and body (and in so doing, inverting ayurvedic theories), served to reassure European readers of the lower strata that Indians occupied.

Medical writing formed part of a greater defensive complex for military and medical commentators. Authors warn of the 'threats' posed by the unclean, uncaring mass of Indians by punctuating their arguments with accusations of outright medical negligence by those practitioners who had not been educated at the government-sponsored colleges. This represented a significant shift from the earlier dependence on and borrowing from Indian medicine. A large number of European medical officers now found the practices of the *vaids* incredibly threatening. Two of the most frequently used terms with reference to native practitioners were 'ignorance' and 'superstition' (most often used in conjunction). Kenneth MacKinnon, in his 1848 medical treatise on Bengal and the North West Provinces, railed against the *vaids* and argued instead for the implementation of two interlinked plans: a more stringent medical police to enforce sanitary regulations and the placement of greater numbers of Indian men who had been educated in the Calcutta Medical College in the *mofussil* areas.[106] Another solution to the supposed backwardness, it was posited, lay in clearly structured, (militarily) organised hospitals and dispensaries of European import. Surgeons hoped that these might pull India out of the morass it had hurled itself into.

Such ideas of medical and moral destitution and destruction reinforced the idea that India, or, more specifically, Indians themselves, presented a threat not only to the Europeans who came into contact with them, but to the stability of Britain's imperial project as a whole. Taken one step further, at the most intimate level, the bodies of Indian women (with their supposedly inherent sexuality) posed an even greater threat, as mentally, morally and physically, they were thought to have enormously destructive potential. City and general 'native' hospitals only began appearing in greater numbers in the 1830s, before which time the only other groups which surgeons were consistently able to observe were prisoners and inmates of the newly-founded insane asylums. Therefore, the observations on the 'threats' India presented to Europeans were most frequently based on the health of soldiers in the army, as this was the easiest group to analyse. Due to the continuing high levels of venereal diseases among the troops, a search for a 'cure' to the problem continued to occupy the minds of surgeons and army administrators alike. Moreover, doctors proclaimed that a large number of the men were rendered permanently invalid by the aggressive and complicated relationship between syphilis and other diseases. As a result, the men were sent before the invaliding board and from there back to London (at a loss to Crown and Company).[107]

The 1831 Bengal Medical Board circular on venereal disease

After the abolition of lock hospitals in Bengal (and clearly concerned about the potential 'explosion' of venereal cases predicted by some surgeons), the Medical Board sent out a detailed circular, requesting their surgeons to furnish answers to 46 questions 'relative to the nature and treatment of the Venereal Disease in India' (See Appendix 2).[108] Circulars such as this were vital to the production of knowledge on subjects of particular importance. It also represented a clear attempt by the military and medical boards to find the most efficient and economically viable way of protecting European soldiers from venereal disease in light of the closure of the lock hospitals. This particular circular was sent out on the orders of the Bengal Medical Board (rather than the government in Calcutta). Once again, the circular and the surgeons' replies to it, challenges the image of a monolithic, carefully coordinated Company state. In their responses to the circular, surgeons spelled out their (often very different) experiences and views on treating venereal diseases in India. Most of these directly contradicted the 'official' line on venereal disease control – which condemned the system of compulsion and forced treatment – put forward by Inspector General of Hospitals Burke, and endorsed officially by Governor-General Bentinck.

What quickly becomes evident when reading through surgeons' replies is the insistence of many that syphilis in India was different from its European counterpart. This allowed surgeons to stress the importance of their specifically Indian medical endeavours and it lent further weight to the notion of racial difference. While many of the respondents declared venereal disease to assume a much milder form in India, the widespread nature of the disease among European soldiers and the implications this had for the maintenance of colonial rule meant that the disease merited urgent consideration.

The replies to this circular, which came in through 1831 and 1832, displayed a broad range of medical experiences across Bengal. These represented an attempt by surgeons to organise a coherent response to the threat of venereal disease in the wake of lock hospital closure. The questions and replies can be categorised into five broad categories: treatments and treatment times; the nature of the disease in different individuals; indigenous methods of treatment and their effectiveness; the difference in the disease between India and Britain; and the suitability of different treatment methods on Indian versus European bodies (See Table 3.1).

The questions in the circular not only suggest the 'threat' venereal disease presented to the robustness of the Company's military, but also imply that the Bengal Medical Board was still amenable to borrowing from 'local' Indian medical practices, should they prove more successful in treating the disease. When viewed together, the replies to the circular present a remarkable pattern. Not only do they reveal the surgeons' preference for mercury-based methods of treatment, but they unwittingly betray the fact that European treatments were strikingly similar to those of their native counterparts. Even while European surgeons attempted to dismiss indigenous practices as dangerous and un-enlightened, it is clear that, by whatever name it was given, or whatever form it took, mercury was the preferred method for both groups of practitioners.

The questions posed in the circular begin generally, with those comparing the success of non-mercurial versus mercurial treatments. However, the circular quickly betrays a concern with cost, in both real terms and in man-days lost, of treating venereal cases among the European soldiery. The circular pressed surgeons for their opinion as to whether the men were, following treatment for venereal disease, wholly fit for duty. It asked specifically if they were ready for '... immediate exertion or exposure to the weather, such as might be expected in the course of active military duty.'[109] The majority of replies to this question, while noting general fitness for duty, stated that the average time spent in treatment was just over one month.[110] A number acknowledged that unpleasant side effects from the mercury could linger for quite some time. These included muscle pains, skin lesions and inflamed tonsils.[111] More often, however, surgeons stressed that it was the extended treatment times which were the greatest concern. When

Table 3.1 [Extract of selected questions] relative to the nature and treatment of the venereal disease in India

With mercury	Without mercury
1. Will have the goodness to state generally the extent of your experience in the treatment of venereal ulcerations on the penis with and without bubo during your residence in India.	The same question without mercury.
5. Describe the treatment both general, and local, usually employed by you for the cure of primary venereal cases.	Ditto.
8. Were the subjects of all such cases fit for immediate exertion or exposure to the inclemencies of the weather, such as might be expected in the course of active military duty.	Ditto.
11. Were any of the subjects of the cases treated by you and how many so injuriously affected by the disease that had they belonged to the Military profession, they must have been invalided.	Ditto.
27. When more than one person has had intercourse with the same diseased woman, are all the resulting affections similar, if otherwise, what differences are observed in primary and secondary symptoms.	
29. Are you of opinion, that the character of the chancre depends on the virus, on the constitution or habit of body, or do you consider that it may be made to assume different characters according to the treatment.	
32. Is there any other variety observable in the persons contracted by promiscuous intercourse, than that which produced Gonorrhoea and chancre.	
35. Do you consider the venereal disease, which prevails in this country more or less severe than that which prevails in Britain.	
36. With reference to the constitution, mode of life, and diseases which prevail among the natives in the different provinces of India, do you consider them more or less fitted for the mercurial mode of treatment than the natives of our own country.	
42. Have you had recourse to fumigation with mercury. What preparation have you used, in what dose, and to what extent both generally and locally.	
43. If you possess authentic information, will you state the general practice in use among the Natives in such cases and the success with which it is attended.	
44. If you possess authentic information will you detail the various modes in which the natives administer mercury.	
45. Are you aware of the local applications to which the natives have recourse in venereal ulcerations, and with what success.	

Source: *Circular Letter to the Superintending Surgeons of Divisions*, 14 October 1831. NAI, Home (Medical).

buboes were present, estimated treatment times doubled and the men could be hospitalised for up to 2 ½ months.[112]

The majority of respondents, while professing that they had little or no real first-hand knowledge of the practices of Hindu or Muslim medical practitioners, were nonetheless happy to offer their opinions based on their perceptions of indigenous treatments. Assistant Surgeon MacGaveston, offering his second-hand knowledge of indigenous medical practices scathingly stated that

> ... the Hindoo Physicians administer mercury and generally in an unmerciful manner or quantity which mode of treatment (i.e. saturating the constitution with mercury) is not followed by the orthodox Hackiems [sic] though the irregular musselmen practitioners also adopt that highly improper and unnecessary practice of continuing its use until the face swells prodigiously and the saliva flows in unmeasurable quantities ...[113]

In contrast to MacGaveston, Assistant Surgeon Finch in Meerut was more critical of what he saw to be the immoral practices of Europeans in India. Stating that the

> ... general mild character of the disease [among Indians] in this country must be ascribed to climate, the modes of life, and diet of the Natives and to the absence of those indiscretions both in diet and drink which lead to the deplorable effects of the disease which we so frequently are witness of more especially among the lowest class of females in Britain.[114]

Once more, we can see how there was still a certain cross-fertilization between Indian and European medical practices in India. Perhaps tellingly, Finch was one of the very few surgeons who admitted approaching *hakims* to discern which methods they utilised in the treatment of venereal disease. Nevertheless, the lingering insistence of a difference between European and Indian bodies was still present in Finch's report.

The circular reflects and reinforces the contemporary medical belief in the ability of 'promiscuous intercourse' to cause syphilis. Question 27 asks, 'When more than one person has had intercourse with the same diseased woman, are all the resulting affections similar, if otherwise, what differences are observed in primary and secondary symptoms?' The responses to this question similarly suggest the formation of a hypothesis that posited that, despite being 'infected' from the same woman, different men displayed diverse symptoms; it was supposed that this was due to each individual's constitution, 'habit of body' and 'to the attention paid to cleanliness at the commencement of the disease.'[115] Assistant Surgeon Finch returned to his preoccupation with class and insisted that the 'character' of venereal sores was made worse by the fact that the soldiers were 'having impure intercourse with the lowest of the sex'.[116] While Finch placed most of the blame on the women's low class and character, Assistant Surgeon George Smith insisted that the men's heavy drinking and exposure to inclement weather were key factors in determining the speed at which secondary symptoms emerged.[117]

Surgeons designated 'mercurial' from 'non-mercurial' cures in a curious manner. A number of practitioners insisted that they preferred 'non-mercurial' treatments, but what these actually entailed were simply lesser doses of mercury, or one which did not produce ptyalism, or excessive salivation, in the patient.[118] Surgeons administered mercury internally and externally. Patients were given pills by mouth, the most common of which was known as the 'blue pill', a mercury compound. These were combined with topical lotions and 'washes' that were rubbed onto the skin. These topical treatments often combined mercury with other mineral, vegetable or plant compounds. Non-mercurial treatments were frequently as dangerous as mercurial ones, containing such things as nitric acid, lead and silver nitrate.[119]

The debate surrounding the use of mercury in venereal treatment was not a new one. In 1820, George Ballingall, a prominent regimental surgeon who had served in India and Java, wrote an essay on the treatment of venereal disease that mocked those arguing against the use of mercury. He noted that '... we are called upon to relinquish an article [i.e. mercury] which we are in the habit of prescribing by grains and scruples, and to substitute one which ... ought to be given in the shape of a pudding or a pie.'[120] Surgeons in India combined these mercurial treatments with plant medicines such as salap[121] and opium as well as making use of such practices like the application of leeches.[122] Once again, such treatments were not dissimilar to those of native practitioners, although often the latter appeared to substitute more varied plant medicines to produce their topical rubs or alternatively, relied on different methods to administer the mercury. However, this dependence on mercury remained central to both European and indigenous practice, despite the continued insistence of European surgeons on the great gap which existed between the two practices. The continued hybridity of European medical practice in India (with regard to venereal disease) is clearly evident in the responses to the circular.

In attempting to distinguish themselves further in the developing medical field, a number of surgeons asserted that the form venereal disease took in Indians was different to that seen in Europeans. This was attributed to a number of different factors. Some surgeons confidently proclaimed that the native 'mode of life' and temperate character rendered them less susceptible to venereal disease; those who caught the disease experienced a much milder form.[123] Surgeon Tytler, responsible for the 50th Regiment Native Infantry, made the rather curious argument that the excessive consumption of rice produced a distinct variant of gonorrhoea. He had come to this conclusion following a series of experiments he conducted on goats while serving in Java.[124]

The range of responses from surgeons across the presidency to the 1831 circular displays not only the range of practices – from the traditional – mercury, the 'black wash' and calomel to the more unusual – leeches, poppy heads and madder root, but also highlights the equally varied biases and beliefs of European medical practitioners. Responses such as these contributed to the growing Anglo-Indian medical 'voice' which sought to distinguish its findings within the European discourse. The circular also represents the military and medical boards' attempt to find the most efficient and economically viable way of protecting European soldiers from venereal disease in light of the closure of the lock hospitals. The circular itself served to consolidate surgeons' knowledge of venereal disease treatment in India and materially contributed to the construction of a unified, common framework for how European surgeons in India viewed venereal disease and its treatment.

Journals and venereal disease

The debates reflected in the 1831 circular were echoed in the medical press across the subcontinent. Medical journals such as the *Calcutta Medical and Physical Society Quarterly Journal*, the *India Journal of Medical and Physical Science*, and the *Madras Quarterly Medical Journal* became an additional forum for surgeons to debate the treatment of venereal diseases across India and consolidate their knowledge. Anglo-Indian medical journals initially reiterated and reinforced current European findings on syphilis with sections that paraphrased or re-printed recent British medical work. These, too, were increasingly infused with moralistic overtones. However, journals quickly moved towards putting forward 'uniquely' Indian experiences of disease control.

In one article published in 1837, Mr Wallace stated that, 'A single day's debauch will cause a complete change in the exantematic sore;

and, on the other hand, a day of abstinence and quietness will restore it to its pristine state.'[125] The same journal featured an article by Mr Cowper, who presented a view common at the time that syphilis and gonorrhoea were in fact the same disease. He confidently asserted that '... every surgeon knows that two men may have connection with the same woman, and one be infected with syphilis and the other with gonorrhoea.'[126] Again, we can see that, through statements such as this, disease transmission theories focused on the body of the woman and were interwoven with social mores introduced in the hope that such threats might scare 'gentlemen' into better behaviour. Cowper went on to assure his readers that the 'habits' and 'temperaments' of an individual determined whether they would exhibit gonorrhoeal or syphilitic symptoms. Those with 'sanguineous' habits were more prone to the gonorrhoeal type, whereas those with a 'lymphatic temperament' were the would-be syphilitics. As 'proof' of this, Cowper noted the case of one unfortunate individual who suffered from 17 separate gonorrhoeal outbreaks while never once displaying any syphilitic sores.[127] Cowper's language reinforces the idea that there was a fluidity of the disease; depending on one's 'temperament', venereal disease could manifest itself in very different ways, or, indeed, not at all.

The anxiety that prompted the Bengal Medical Board to solicit knowledge via the 1831 circular repeatedly struck European surgeons across India. As will be discussed in Chapter 5, following the abolition of the lock hospitals in Bombay and Madras in 1835, surgeons not only took to lobbying their medical boards for reinstatement, but quickly made use of the newly established medical journals published by the professional societies to argue their case. Here again, we can detect the ways in which European surgeons in India not only demanded the return of the lock hospital system, but sought to stress the specific importance of the medical knowledge they offered to their peers in Europe. In 1837, the 'Surgery' section of the *Calcutta Medical and Physical Society Quarterly Journal* dedicated two (of its four) articles to the treatment of syphilis.[128] Assistant Surgeon Clark, of HM 13th Light Dragoons, wrote in 1839 in the *Madras Quarterly Medical Journal* that 'of the common and every day diseases of our frail mortality, or still more frail morality, none claims more attention than syphilis.'[129] Clark confidently warned that syphilis was '... produced by an act of generation in promiscuous intercourse [and] it carries along with it into the constitution a generating power, and impregnates the previously latent seeds of disease.'[130] Observers proclaimed that not only did syphilis directly break down the constitution of the sufferer, it could activate diseases the carrier had been previously unaware of, thus wreaking havoc on their system in devastating and unpredictable ways. Moreover, they suggested that a woman might convey disease without herself actually coming down with symptoms.[131] In a similar manner, in a series of lectures given in London in 1838–1839 on the subject of venereal disease, Frederic Skey suggested that the 'poison' of venereal disease could develop through the action of an 'irritant'. Such potential irritants ranged from leucorrhoea to menstrual fluid, to what Skey deemed 'mechanical irritation.'[132] Thus, an unsuspecting (or unsuspected) woman could provoke or unleash the venereal 'poison' in her partner, even, Skey seemed to insinuate, if she was a virgin. Apart from his dedication to alerting his fellow medical observers to the pernicious effects of the disease, Clark's phraseology suggests an attempt to feminise the disease and situate blame for the disease in the supposed originator of such illicit intercourse – the 'loose woman'. Drawing heavily upon moral arguments, Clark blamed what he perceived to be the degraded state of society in India for increased levels of disease. 'Prostitution prevails to a great extent at such periods [of famine], and the state of debility induced, appears to be favourable to ulcerations and discharges from the sexual organs of the female.'[133]

Clark used his article as an opportunity to rail against the abolition of the lock hospitals. He argued directly against the findings of the late William Burke. Governor-General Bentinck saw the system as a complete waste of precious Company money and quickly adopted Burke's report. Yet surgeons did not accept Bentinck's decision to abolish the system lying down. They employed every available weapon in their arsenal to fight this ruling. Burke, Clarke asserted, failed to take into account the fact of the greater levels of virulence of syphilis at stations where lock hospitals had not been present. Continuing on a pseudo-moralistic line, Clark argued that syphilis was a disease of the poor, and, as such, it was the *duty* of the government to care for its sufferers by the provision of such 'humane' institutions as the lock hospital. Clark's argument perfectly highlights the belief that poverty is one of the conditions that promoted the generation of venereal disease. This, along with the 'freedom' given to the 'modern' woman, was often enough to render her a diseased mess. The newly important ideals of cleanliness, femininity and motherhood are thrown into sharp contrast with its 'other', the prostitute:

> ... instead of being, *as intended,* a fruitful vessel (garden where the seed of man is planted and nourished,) she becomes sterile and unclean, a rank and loathsome ditch, in which disease in place of children, is generated.[134]

Moreover, certain sexual 'problems' were thought to be more prevalent in warm climates. Even, it was proposed, 'upright' and 'virtuous' women had the potential to become conduits for disease. Harmless secretions took on sinister overtones and were thought to have the ability to mutate through 'over use' or 'abuse' to become harmful new diseases. To compound the problem of these hot-weather latent sexual threats, many observers believed that debauchery was also generally more common in hot climates. Thus, surgeons and commanding officers proposed lock hospitals again as the last hope against a number of otherwise insurmountable venereal obstacles. John Murray, the Deputy Inspector General of Hospitals in Madras, also petitioned for the return of the lock hospital system in 1839. He, like Clarke, felt Burke's recommendations were flawed and had led to a catastrophic failure to contain the 'evils' presented by venereal disease. Again, blame for the disease quickly shifted to women. Murray wrote almost pleadingly that, 'It cannot be doubted that the source of the evil is in the diseased Females being allowed almost unrestrained intercourse with the soldiers.'[135] These fifth-columnists 'lurking' within cantonment lines were felt to merit the same type of forceful discipline and control suggested by the earlier writers on the 'animalistic' and 'uncivilised' tribes and peoples. Following the abolition of the lock hospital system, surgeons and commanding officers, through the medical boards, professional societies, journal and essays, moved in a synchronised way to push for the return of the lock hospital. The 1831 Circular and the publication of numerous articles on venereal disease in the Indian medical press represented one of the first concerted efforts by surgeons to form a unified approach – both medically and politically – to the problem of venereal disease in India.

Conclusion

While scholarly attention has previously focused on the actions of liberals and utilitarians in India as they negotiated their way through the 'Age of Reform', the changes taking place in the fields of medicine and science, fuelled by the demands of the military, were often more significant. The observations and interpretations of surgeons on the health of the European soldiery was a way in which they could reinforce the security of the Company state itself. Possible 'threats' to the stability of the Company, medical and otherwise, were personified, highlighted and reiterated until they became clear social and racial stereotypes. This same process transformed the perception of different groups that ranged from the 'effeminate Bengali' to the supposedly sexually aggressive prostitute. As both a physical and moral threat, the latter provoked much stronger reactions among military and medical men.

Surgeons and observers rushed forward to provide a host of theories covering everything from venereal treatments to comparative anatomy. Increasingly, their 'evidence' stressed the supposed social, sexual and biological differences between Europeans and Indians. These ideas spread widely across India, due in part to the medical and scientific networks which medical boards, colleges and societies provided. Surgeons were quick to take advantage of the opportunities which medical circulars and journals provided to boldly put forward their own 'unique' theories and demands. The increased formalisation of medicine and the growth of the medical press enabled surgeons to become a more coherent pressure group. When it came to venereal disease, this lobby spoke with one concerted voice. The ideas produced during this period had far-reaching consequences, not just for the development of science, but for patterns of imperial rule. As this evidence emerged in response to the perceived 'threats' facing both the military and the stability of colonial rule, they constructed an image of India and Indians which was heavily refracted through the prism of military and medical concerns. The theories that emerged from this crucible suggested clear biological differences between Indian and European bodies. These racialised, sexualised bodies would remain prominent in the colonial imaginings of India well into the twentieth century.

Plate 1 A nautch at Hindu Rao's House © The British Library Board Add.Or.4684

TOM RAW GETS INTRODUCED TO HIS COLONEL.

Plate 2 Tom Raw gets introduced to his colonel

Plate 3 Qui Hi pays a nocturnal visit to Dungaree

4
The Body of the Soldier and Space of the Cantonment

While preceding chapters have focused on the attempts to control the bodies of Indian women in and around the military cantonments, it is important not to ignore the ways in which government and the military imposed discipline on and around the body of the European soldier himself. This chapter examines the methods and levels of control – both which existed and which were attempted in the cantonment. These ranged from regulations enacted to order the physical space of the cantonment, to calls for a more direct control over the bodies of the soldiers themselves, or, more commonly, the numerous others who occupied cantonment space. Crucially for this argument, moral and medical concerns were of critical importance in moulding this ordering.

Although the fault for venereal infection was increasingly framed in gendered terms, there were other 'moral' ailments stalking the cantonments that were harder to blame solely on 'dangerous' women. Among the most pernicious of these was the high level of drunkenness among the European soldiery. Concern among surgeons and commanding officers with the sheer volume of drink-related cases filling hospital beds as well as the number of men populating the invalid list was overwhelming.

As suggested in earlier chapters, each of the three Presidency armies possessed different characteristics. Similarly, there were key differences in the recruiting patterns and composition of Crown and Company troops. Recruits to the Company's rank-and-file were generally seen to be of a 'higher grade': these men enjoyed better pay, prospects and conditions of service than their Crown counterparts.[1] However, within warnings about the perils of drink among the European soldiery in India, such distinctions vanished. Similarly, a soldier's origin (be it Dundee, Belfast or Leeds) was largely irrelevant when it came to portrayals of the men's social or 'leisure' activities and in particular, in attitudes towards drink. In these accounts, the men's class background proved a great levelling factor. Drinks were themselves starkly divided by class lines, with those favoured by the middle- and upper- classes (claret and beer) portrayed as acceptable (and even 'wholesome') and those drunk primarily by the working classes (spirits and arrack) deemed dangerous and loathsome. Military and medical observers on the health of the troops in India blamed these drinks, and by extension, the men themselves, for a precarious state which saw a steady stream of hospital admissions due to intoxicants.[2] The men, largely recruited from working-class backgrounds, were portrayed as resolutely low and degraded, while possessing the brute strength necessary for the maintenance of British rule in India. This perception of the men was a critical element in the construction of one of the tropes of colonial masculinity. This construction rationalised any aggressive, violent behaviour from a European soldier as 'natural', while 'effeminate' behaviour or homosexual activity was seen to strip the men's vitality and power.

The various ways in which the military and government imposed order on the cantonment had broader implications for European understandings of India's inhabitants and the shaping of the empire itself. This chapter explores how intemperance among the soldiers prompted demands from medical and military officials to re-define the occupation of cantonment space. Anxiety about the intemperance and misbehaviour of the men gave rise to a raft of cantonment regulations that not only imposed a punitive regime on those living and working in and around the cantonments, but prompted an extension of military space.

Somewhat ironically, it was often the case that the soldiers themselves were blamed only indirectly for their drunken behaviour. Assumptions about the men's class and character meant that most officers saw the soldier as a creature of animal needs. Although there has been significant work on the perceived malleability of the soldier in the eighteenth century,[3] this chapter examines how this perception informed attempts to regulate temperance and moulded the disciplinary regime in the cantonment. It argues that, much like the ways in which military and medical demands shaped European understandings of Indian women, the idea of 'race' and Indian society more broadly, so too were these notions critical in reinforcing representations of the 'typical' British soldier. Through an examination of the other groups blamed for the soldier's misbehaviour, namely 'disorderly' European women and illicit liquor vendors in and around the cantonment we can see the ways in which these constructions were reinforced. Finally, it explores the actual punishments imposed for misdemeanours committed in the cantonment, from drunken behaviour to sodomy and rape, in order to show how such class-based understandings, and a preoccupation with the 'masculinity' of the European soldiery, shaped penal regimes.

Intemperance and the soldier

In his 1858 series on maintaining the health of the European soldiery, Norman Chevers, then a surgeon in the Bengal Medical Service (who would go on to produce a number of medical manuals for India), began a section dedicated to temperance with the warning that it was impossible to really convey how dire the situation of drunkenness among the European troops in India truly was.[4] One anonymous author, writing in the *Calcutta Review* in 1858, solemnly assured his readers that, '... nowhere does drinking flourish more rampant [sic] than in India, where it is a principal source of much of the sickness and mortality which exists.'[5] Indeed, when compared to similar statistics of the army in the United Kingdom, the situation in India did appear much more dramatic. In the late 1830s, admissions for delirium tremens,[6] were 27 out of an overall troop strength of 44,611 for the 'home' army. This figure was unfavourably compared with admissions for the Bengal army, where there were 672 admissions and an overall strength of 36,286.[7] Like the surgeons who warned that the 'variety' of venereal disease present in India was more widespread and pernicious than its European counterpart, the author of 'The European Soldier in India' cautioned that the liquor supplied in India was not only more readily available than in England, but was itself of a more noxious variety.[8] This idea – that the drink most commonly consumed by the soldiers was toxic – quickly became a theme that ran though proposals to remedy the 'drink problem' among European soldiers in India.

As the perceived fuel for much of the men's misconduct, drink was also seen to open the door to more serious ills, from delirium tremens to death by cholera. Contesting the view that the area around Kaira[9] was generally more unhealthy than other parts of Gujarat, Mr Gibson,[10] assured an audience of his peers at the Bombay Medical and Physical Society that the poor state of the European troops stationed at Kaira was solely due to the men's fondness for local liquors – arrack and a 'Mhowra' spirit. The two demon drinks, Gibson believed, rendered the men particularly susceptible to malaria.[11] Alcohol shared this dubious honour of being a 'gateway' to deeper ills with venereal disease. Indeed, medical observers suggested that had the men not been such inveterate drunkards, they might be more attuned to the venereal threats posed by illicit prostitutes 'lurking' in and around cantonments. The link between 'vice' and illness was made explicit in an 1855 article titled 'Soldiers; their Morality and Mortality' which appeared in the *Bombay Quarterly Review*.[12] Most admissions to hospital, the author proclaimed, resulted from two causes – 'vagrant amours' and drunkenness.[13] It is no coincidence that the perception of both was that they were not only preventable, but driven by human weakness. The article itself was an essay-style review of eight contemporary works on the subject of temperance and the health of the soldiery. Its author attempted to display, through the careful use of statistics, how vice directly led soldiers to ill health, poor performance and bad behaviour. The author unflinchingly declared that there was a sure link between morality and mortality, noting that the, '... laxity of morals has a tendency to shorten the period of human life, and purity has a corresponding tendency to longevity.'[14]

The Company's attitude towards intemperance and its associated health concerns was strikingly similar to that taken with regard to venereal disease. Like venereal disease, medical and military reporting approached drink with two fundamental assumptions: first, that it was one of the primary causes of inefficiency in the corps and second, that the soldier was naturally inclined to drunkenness. The occasions on which drunkenness *actually* prevented soldiers from actively engaging with an enemy were rarely itemised. Instead, this vice was presented as endemic among the European corps in India.[15] As such, what is important is how this description was framed and repeated *ad nauseam*. The argument about the drunken soldiery, in a sense, became a mantra to elicit support for various demands from the government in Calcutta or London. Medical and military concerns combined to create a new type of morality imposed on cantonments. Furthermore, like syphilis, intemperance was seen not as a problem of the individual soldier and his self-control (or lack thereof), but of the soldiery as a whole. According to this rhetoric, the lure towards intemperance and bad behaviour came from without. It dismissed the agency of the individual soldier in actively seeking out drink (illicit or otherwise). Instead, given his 'natural' propensity to drink, attempts to regulate drink targeted those who 'encouraged' him to do so. The individual soldier was not blamed for his intoxication, unless, of course, it went beyond what was considered 'normal'. For most officials, drunkenness was part of the composition of a British soldier. As will be discussed, this very belief is reflected in the punishments (or lack thereof) inflicted on men for drunkenness as well as in the laws and regulations drawn up to sanitise cantonments.

Military and medical descriptions of the European soldier

In his 1846 account, John Macmullen, formerly Staff Sergeant of the 13th Light Infantry, recalled landing at Calcutta following the long sea voyage from England. On returning to his quarters after a wander through town, he came upon the men of his regiment, completely dissipated: in various states of intoxication, breaking up their *charpoys,*[16] and engaged in pitched battles with each other. 'Several,' Macmullen reported disdainfully, 'completely showed the cloven foot.'[17] More worryingly, Macmullen admitted that the 'non-commissioned officers possessed no control whatever over the men, by whom they were thoroughly despised.'[18] The scene Macmullen described was, by many accounts, a common occurrence and this view of the men, as aggressive, intoxicated louts, was a widely held one. No less a figure than Arthur Wellesley, years after his return from India and at the time of the missive, serving as Prime Minister, wrote an impassioned memorandum objecting to the proposed abolition of corporal punishment for European troops in India. He argued that given the degraded state of the men whom he claimed filled the ranks of the European soldiery, physical violence or the threat of violence, was the only way to hold them in check.[19] The man who enlisted in the British army was, he insisted,

> ... the most drunken and probably the worst man of the trade or profession to which he belongs or of the village or town in which he lives. There is not one in an hundred of them who, when enlisted, ought not to be put in the second or degraded class of any society or body into which they may be introduced; and they can be brought to be fit for what is to be called the first class only by discipline...[20]

The author of 'Soldiers; their Morality and Mortality' went even further in his analysis. He divided recruits to the army into three groups. The first, and largest of the classes, was composed of men, 'whose follies and vices have alienated them from their friends, and rendered them unfit for the duties of civil life'.[21] Men in the second category were those who, by virtue of their poverty, entered the army to provide for their subsistence. The third category, which the author clearly viewed as the most 'virtuous', was composed of men who chose soldiering as a profession due to their own personal spirit of adventure. Class assumptions and prejudices were at their most blunt when discussing the health of British soldiers. By this standard, those who chose soldiering out of financial need were less noble than those who had the freedom to 'choose' a life of adventure.[22] The largest group of soldiers, by his estimation, appeared little better than common criminals. These civil misfits, it is implied, were dumped into the army as there was nowhere else safe to put them (besides, perhaps, the prison). It was reasonable, then, to assume that such vice-filled men would prove dangerous when given access to drink.

Anglo-Indian observers on the problem of drink were careful to assure their readers that in the *respectable* circles of society, there was a general mood of temperance. James Johnson suggested that this healthy physical state, combined with good manners, was *essential* to preserve Britain's empire. The drunkard, Johnson reasoned, not only destroyed his own health but was responsible for, '... deteriorating the European character in the eyes of those natives, whom it was desirable at all times to impress with a deep sense of our superiority.'[23] Monier Williams, Surveyor-General in Bombay (and father of Sanskritist Monier Monier-Williams), concurred with this assertion, lamenting in 1823 that through drunkenness,

> ... the English soldier exhibits to the natives of India, a disgusting specimen of the European character, – the Christian character, and even of human nature itself. This abominable vice of drunkenness, which brings such dreadful evils upon themselves, and such disgrace upon our national character and religion, among our native subjects, is almost universal among the English soldiery.[24]

Observers frequently echoed this vision of the low-class soldier, with another confidently assuring readers that 'The ratio of deaths among the *higher* classes of Europeans is less by eighty per cent than among the troops.'[25]

Contradicting this glowing vision of the higher classes was a statement made by Chevers in one of his reports that officers, from the very earliest days of European occupation in the country, were prone to intemperance. Chevers cited Mr T Baker who noted that in past times, one-third of officers died 'from the effects of snipe shooting and intemperance.'[26] Whether or not these deaths resulted from an unhappy combination of the two was left to the reader's imagination. Lest we be seduced into believing that the officer class was free from this 'vice', it should be remembered that officers were as fond of swilling Hodgson's pale ale (in between bouts of shooting and cricket) as soldiers were of the less-expensive offerings to be found in the bazaar. However, the 'higher classes', with their more regular habits and careful choice of drink, were, in this instance, thought better able to withstand the rigours of the climate thanks to their good manners and upright behaviour. However, most observers on drink and the soldiery simply conveniently overlooked the differences in lifestyle that encouraged the good health of the well-to-do, such as superior diet, housing and their reduced exposure to both sun and violent conflict.

Chevers used his reports to warn of the possible threats contained in allowing this torrent of intemperance to continue. These consequences not only included impaired health, but were, he suggested, bound to much greater imperial concerns. Apart from the Dutch and the Portuguese (who he similarly regarded as excessive drinkers), he noted that the other '... continental nations of Europe ... are vastly our superiors, and have abundantly proved their greater fitness to bear exposure to the climates of the tropics, both in the East and in the West.'[27] The French in Pondicherry, he added, were almost entirely free of hepatic complaints, subsisting as they did, on light wines, vegetables and curry.[28] No doubt Chevers hoped to provoke his readers into action with the perceived threat to imperial and national prestige and security, timed as it was so soon after the Crimean war and the 1857 Rebellion.

While the image of the European soldier was somewhat rehabilitated in the eyes of the British public in the years following the 1857 Rebellion, most surgeons and commanding officers (and indeed those like Wellesley who peopled the highest ranks) remained disparaging in their remarks about the quality – moral and physical – of the European trooper. Even the 'Mutiny Diary' of William Russell, War Correspondent for the *Times*, betrayed this disparaging view of the soldiery. In Russell's depressing account of the sack and pillage of Lucknow by European troops 'drunk with plunder' he observed that, 'Discipline may hold soldiers together till the fight is won; but it assuredly does not exist for a moment after an assault has been delivered, or a storm has taken place.'[29]

However, as we have seen, any attempt to prevent the men's 'natural' urges, it was feared, could provoke unwanted consequences, either pushing them towards homosexuality (and therefore 'emasculation') on the one extreme, or an explosion of pent-up 'passions' (as Macmullen described it). While the Company desperately needed European soldiers to reinforce and maintain control, these soldiers struck fear into the hearts of their own officers. This disquietude influenced the measures that the colonial state enacted to preserve the men's health. Thus, while Sara Suleri has observed that an overwhelming fear complicated the Anglo-Indian narrative – that of India and its 'incomprehensible' traditions – so too were decisions regarding European soldiers spiked with a similar panicked sense of incomprehension.[30]

Observers presented two pictures of the soldier which sat rather uneasily with each other – that of the 'naturally' intemperate, low-class ruffian, and, conversely of the naïve victim of outside forces who lured him into a life of drink and misbehaviour. Although at first glance these two visions appear to be contradictory, when we consider that the theme common to both was that soldiers were, in effect, malleable, the two become more complementary. As Foucault has argued, by the eighteenth century, '... the soldier has become something that can be made; out of a formless clay ... the machine required can be constructed'.[31] As such, the Company and Crown both harboured the belief that they could mould 'low-class' men into efficient, masculine soldiers. However, such pliancy also meant that the men were equally exposed to the whims of those bent on less salubrious and more salacious habits. Threats to the men's health could arise from any number of sources, but when it came to drink, those thought to be the most dangerous were the illicit liquor sellers 'lurking' around the bazaars and the soldier's wives, secretly selling liquor from their quarters. This set of assumptions regarding both the men and the various threats to their health definitively shaped the construction of drink-related regulations.

Canteen and cantonment: medical theories and proposals for military spaces

Those seeking to reform the Company's approach to intoxicating substances did not lack medical reports to support their allegations. As in the struggle against venereal disease, the war on drink was waged by an alliance of medical men, commanding officers and (after 1813), missionaries. However, by the time the latter established themselves across India, this coalition of medical and army men had already substantively shaped the debate on the issue. The Age of Reform is usually described as one powered by an alliance of free traders, utilitarians, liberals and missionaries. However (as discussed in Chapter 3), equally important were a host of other characters; in this case, surgeons and commanding officers who, once again, dictated their demands to preserve the men's health which, in turn, were carried out by government. Surgeons and commanding officers successfully argued that drunkenness in India was more dangerous to the constitution than it was in Europe. This idea appears to have followed on from earlier assumptions that linked climate, excess and disease. Before the nineteenth century, European observers on 'health' in India sought to project the idea that, given certain precautions were taken in one's lifestyle, India was a perfectly healthy place in which to live.[32] However, these observers warned that the heat compounded the many excesses common to the lifestyles of Europeans in India.[33] Habits of drink were particularly threatening due to the potential alcohol had to excite and upset bile production, as well as affecting the liver and stomach more generally.

In his 1813 tract, *The Influence of Tropical Climates on European Constitutions,* James Johnson asserted that the excitement to one's system caused by excessive 'debauchery' (here referring to both food and drink), was followed by a general torpor in the liver.[34] This sluggishness in turn obstructed the production of healthy bile that then encouraged fever, dysentery and hepatitis. To prove his point, Johnson pointed to an example of Europeans situated on the Coromandel Coast, the broad, coastal plain of south India. Here, he asserted, Europeans were temperate and sensible and therefore, the instances of hepatitic problems were lessened. This, he noted, contrasted with soldiers and sailors

> ... who, to the stimulating effects of the climate, add inebriety, too much food, or ill-timed exercise, then the biliary secretion and perspiration are so hurried and augmented, and the vessels so debilitated, that the smallest atmospherical vicissitude becomes dangerous.[35]

Johnson urged his readers against this excess, which in his mind was detrimental not only to health, but also to the very stability of European society and control in India.

Medical topographical reports, which multiplied from the 1820s, reinforced earlier warnings about intemperance. These reports analysed levels of health at the various European stations across India. The reports' authors envisioned a range of solutions for countering health problems on the subcontinent. The author of the 1827 report on Bengal noted that:

> Those who are temperate, take moderate exercise, attend to keeping their body open and mind at ease, enjoy good health in almost all these stations in Bengal, as is proved by the inferior degree of sickness which occurs among officers, or persons in Civil Life ...[36]

On the other hand, the report blamed soldiers stationed in Meerut for their 'want of care and prudence', in their reckless attitude not only to venereal disease, but to alcoholic over-indulgence.[37] Reiterating Johnson's theory, he warned that drunkenness paralysed the ducts, obstructing the secretion of healthy bile as well as weakening stomachs already debilitated by India's climate.[38] The author asserted that this dual threat led to liver disorders which, combined with the heat, greatly increased levels of disease.[39] In addition, he disdainfully dismissed the suggestion that liquor be given to soldiers to mitigate the effects of strenuous exercise. To prove the fallacy of this theory, he pointed to the comparatively good health of the native troops, noting that they were subjected to the same (if not more vigorous) exercise as their European counterparts, yet never drank at all.

The prevailing logic argued that those younger soldiers with only a few years' service in the country posed the highest risk for both bad behaviour and 'tropical' ailments. Surgeon Ainslie of the Medical Board suggested a plan for the health of the newly arrived troops that involved a period of 'seasoning'.[40] The idea of 'seasoning' was not new. In 1773, East India Company surgeon John Clark suggested that newly arrived troops be sent to the Coromandel Coast for 12 months to allow them to acclimatise gradually.[41] In these suggestions we see the nucleus of the idea behind later sanatoriums and hill stations. In a manner similar to Clark, Ainslie's plan proposed the creation of a station for the newly arrived at an elevated location close to the sea, properties felt to be essential prerequisites for good health in India. Situated at a distance from arrack and toddy shops, this site would allow the men to adjust to temperature, food and lifestyle over a period of eight months. Ainslie's plan restricted alcohol allowances and prevented young men from 'wandering' outside of the cantonment in search of women. However, Ainslie's proposals (like Clark's before him) ultimately came to naught, as the Medical Board (while supportive of aspects of the plan) ultimately failed to approve it – in all likelihood due to its proposed expense. Instead, both Company and Crown adopted an informal approach, although it seems clear that the 'seasoned' soldier was more highly valued than his griffin counterpart. One soldier, Gunner William Eggleston, had only been in India for four years when he wrote to his sister that, 'When I joined the troop there was 109 men in it, it is now equally strong but only 27 of the same men remain. One seasoned man is worth 12 recruits so you can guess my value.'[42]

The call for more careful attention to be paid to the recruitment and early stages of a soldier's career in India did not disappear. Norman Chevers, in his 1858 series on the health of soldiers, set out step-by-step instructions that he deemed expedient and necessary in order to reduce illness among the European troops in India.[43] These began with the proper selection of recruits and proceeded through the appropriate activities with which to occupy a soldier's mind and time on a daily basis. For Chevers, like Ainslie, optimum health in India was attainable only if military and medical authorities would pay stricter attention to the careful moulding and guidance of young recruits.

Many observers on the problem of drink within the army railed against the canteen system. By this system, an alcohol ration formed part of the men's daily mess. The first modern European-style distillery in India opened in Cawnpore in 1805 to provide rum supplies to the army's European troops.[44] The daily ration for each European soldier was one dram of either arrack, rum or other 'fiery spirit'.[45] During the rainy season or while on field duty this increased to two drams daily.[46] In addition, a soldier could buy a further dram at the cantonment's licensed canteen or liquor shop. It seems likely that the policy of daily liquor allowances grew out of an earlier notion that a 'stimulating' drink could counter the effects of India's oppressive climate on the European constitution. Although this idea fell from favour by the early nineteenth century, the daily ration remained. Monier Williams blamed the canteen policies of both the Company and Crown for fostering the habit of excessive drink. Williams argued that through a system that promoted daily group drinking, soldiers who, on first joining had been adverse to the taste of alcohol, simply grew accustomed to it and the habit of consumption. To remedy this, he suggested that the alcohol allowance be cut entirely, which, he admitted, might result in a few short-term minor mutinies, but, with persistence (and the replacement of tea for liquor), a happy outcome would prevail.[47]

Henry Piddington's 1839 pamphlet, *A Letter to the European Soldiers in India, on the Substitution of Coffee for Spirituous Liquors* proposed a slightly different, caffeinated solution to the problem (as suggested by the title of his pamphlet).[48] He expressed his deep conviction that, were the troops shown how to make a *proper* cup of coffee, the problem of intemperance among them would be greatly reduced. Piddington suggested that the high level of drink in India was due to the anxiety many soldiers felt about their situation. He reasoned that the men drunk grog to make their minds more 'comfortable' and the properties of a good cup of coffee could offer the same relief.[49] Piddington asserted that a soldier could drink 12 cups of coffee [sic] without feeling 'tipsy' and still wake up the next morning without the '... sickness and horrors to which persons who drink are subject'.[50] Coffee and tea became more widely available in regimental mess as the century progressed, but unfortunately for Piddington, this fact did not appear to detract from the men's fondness for alcohol.

In a similar manner, Deputy Lieutenant-General Henry Havelock implored that the ration be eliminated.[51] He boasted that on their entry to Jellalabad[52] in 1842 his garrison gained one-third in 'manual exertion' as a result of the soldiers of the 13th Light Infantry having no spirit rations (the commissariat's supply having been lost on the journey). Havelock proudly noted that instead, every hand was 'constantly employed with the shovel and pickaxe.'[53] The East India Company's Court of Directors did not heed these warnings. But whether this was due to the fear of revolt amongst the men, the perceived expense or alternative financial reasons, is unclear. Douglas Peers argues for the latter, noting that the government and Company not only profited directly from the sale of liquor at the canteens, but gained substantial additional income from the duties imposed on local liquor production.[54] Hence, attempts at moderating the ill-effects of drink sought to promote the consumption of alcohol that was less debilitating and in the process attacked the Company's 'illegal' rivals in the drink monopoly. For any plan to win the approval of both government and military, it needed to enable change without compromising the masculinity and fighting ability of the soldiery.

Alterations to the canteen system *were* introduced from the late 1830s, though these were not as extensive as many surgeons might have hoped (and Piddington's caffeinated utopia never came to fruition). While temperance societies and a few observers focused on the need to reform the men's behaviour, the government saw the greatest danger (with regard to intemperance) lying elsewhere. This menace took the form of illegal liquor vendors in both the bazaar and cantonment. The illicit sale of liquor threatened the government on a number of counts. Not only were the spirits unregulated and often adulterated with more noxious drugs, but as suggested, their sale undercut the government-regulated liquor prices (and therefore threatened not only the profits garnered from the monopoly they retained on liquor sale in the canteens, but also the duties collected from licensed vendors). By this argument, ill health and intemperance arose when the men wandered *outside* of the cantonment walls in search of both liquor and women. Those who supported this thinking argued that grog, toddy and spirit shops sold the most poisonous and threatening of home-made liquors, laced with a cocktail of drugs and hallucinogens. Norman Chevers recounted tales of Flag Street in Calcutta, a street in the *lal bazaar* known for punch houses and debauchery, which got its name from the flags and banners draped across the street that advertised these insalubrious wares.[55] The liquor available to soldiers in this area, he insisted, was the most dangerous variety available, a rice-based spirit drugged with '... *datura, cocculus indicus, gungah* or ... other narcotic[s]'.[56] Chevers described Flag Street as an area choked with a lair of wide, open sewers, around which these clandestine warrens of sin were located. He assured his readers that entry to most punch houses came via a board laid over the open sewer. He recounted a story about one house in particular, which was linked to a number of sudden deaths. In this house (which was very popular with sailors), men who became incapacitated were taken in to the back room and laid out across a plank on the floor to benefit from the cooler air. Unfortunately, it was the rush of the open sewer which ran below which produced this coolness. This unholy waft infected a number of the men with cholera or suffocated them with sewer fumes.[57]

Chevers proposed a redevelopment of Flag Street based around contemporary medical thinking. Through the combined might of the military, government, local merchants, the Calcutta Municipal Commission and, of course, European medicine, he felt confident that such a plan would 'sweep away the evils' of Flag Street.[58] Based on its drainage and proximity to both Fort and shipping, officials would choose a new spot for the regimental bazaar. Next, he proposed the construction of a proper sewer network for the area; the *well-behaved* amongst the Flag Street publicans could apply for permission to resettle in the new bazaar. Close to this 'utopia' (and to complete it), Chevers proposed the establishment of a lock hospital.[59] Earlier regulations had attempted to draw a moral and administrative line in the dirt regarding the boundaries of cantonments. These projected the idea that moral and medical dangers lay *outside* this boundary. Chevers' solution to this problem, therefore, was to integrate the outlying city into the regulatory fabric of the cantonment. The implications of a plan such as this were immense. If seen to fruition, it would have combined control over the individual bodies of soldiers as well as the physical space he traversed. This 'space' had the potential to cover much of British India.

However, it is unlikely that such micromanagement of the troops would have been approved. Judicious as Chevers' plan was thought to be, the costs involved in his proposals were high. After the massive expenditure involved in suppressing the 1857 revolts, and the further cost of maintaining greatly increased numbers of European troops in India, Chevers' carefully thought-out plans for everything from on-board ship sanitation to proper time and place for drilling, were simply too expensive and complex for the government to contemplate.

Ordering the cantonment: military and government regulations

Given these dire warnings, what actions did military and government officials actually take to combat the problem of drink in the army? Explicit in the plans proposed by medical men and commanding officers for ordering the cantonment, and implicit in the rules and regulations governing cantonment land, was a sense of who was believed to be at fault for exciting disorder in the stations. Regulations which aimed at curtailing the soldier's misbehaviour sought to control those around him as well as the actual physical space of the cantonment and surrounding land. This very fact adds a new dimension to current debates about colonial governmentality.[60] Indeed, while the colonial state expanded its framework for surveillance and discipline over the Indian population, in this case, the intended target for management (or protection) was its own, supposedly ungovernable, soldiers. Cantonment regulations did not suggest that the soldier was himself blameless, but placed a greater degree of the burden on others – the unscrupulous wife, the shadowy illicit arrack seller or the (overly) seasoned Old India Hand who, it was asserted, fostered a culture of heavy drinking and uncivilised behaviour. As will be seen, relatively straightforward demands to curb the behaviour of European soldiers developed into much more complex calls to push out the boundaries of military-controlled space, all on the supposed grounds of health and morality. This transformation involved a complex series of understandings – that of the soldiers themselves, of European women in India, and of those Indians (excluding sepoys) in the service of the army, who held positions from translators to dhobis. Later cantonment acts (most notably Act XXII of 1864), would combine military and medical imperatives thought essential to maintain a healthy corps. Piece by piece, these earlier regulations established a normative view of what cantonments would look like; they dictated how their internal economy should be governed, and sought to regulate all military space.

In March 1809, the Government at Fort William passed Regulation III, which sought to fix the physical limits of cantonments as well as more clearly define the respective powers of the civil and military authorities in carrying out their duties in the bazaar.[61] By this regulation, the limits of all cantonments (where more than half a battalion were quartered) were set through the joint decision of the commanding officer and the magistrate at each. The commanding officer and magistrate sent detailed reports on the specific limits to government. This regulation made clear the decision that the responsibility for maintaining the peace and preventing 'public crimes' within cantonments and military bazaars lay with the officer commanding the troops at each. This decision was hugely significant in shaping the understanding and management of cantonment land. Nicholas Dirks has argued the multiple, intricate ways in which the Company asserted its own sovereign position from the mid-seventeenth century.[62] This 'right' to rule often masked the aggressive violence and corruption of Company rule. In the case of the cantonments, this aggression was only very thinly disguised. The Company's physical control over cantonment space was one of the most blatant of these means. Through a clear demarcation of military land and by ultimately assigning punitive control for crimes committed within those bounds to a military official, the Government at Fort William made clear its partiality toward military-controlled, over civil, space. Cantonments were the areas where some of the most explicit, bald and direct control over the local Indian population existed. Regulation III and those that followed were, in effect, a preview of how the military would dictate later demands for the direct control of land across India.

To bolster Regulation III, further additions came in 1810 through Regulation XX.[63] What Regulation III rather tentatively begun, Regulation XX boldly and aggressively pushed forward. By this act, all persons *associated* with military establishments were subject to military law, in effect extending the Articles of War[64] to every Indian serving the troops, either as retainers in camp or in the field. Furthermore, whereas Regulation III had left both the prevention and prosecution of petty crimes to the magistrate, it was now expressly stated that any person serving the army and in receipt of public pay was now subject to trial by court martial for *any* offence ranging from petty theft to murder.[65] This provision was even extended to those persons not directly in the public pay, but nevertheless serving the army (such as the 'menial servants of officers').[66] Through this Regulation, the punitive power of the military in India expanded monumentally. The regulation firmly and swiftly established legal and penal control over any and all of the army's legions of servants.

After the enactment of these regulations, demands for additional 'moral' control over the cantonment rapidly escalated throughout the 1820s and 1830s. Moreover, commanding officers sought even greater (and more absolute) personal control over their cantonments. Correspondence in 1826 between the Commissary General, Mr Morison and the Chief Secretary, Mr Hill notes that commanding officers and bazaar police sought the power to inflict swift corporal punishment on camp followers and retainers who had either committed a petty offence or conducted a breach of 'good order' *without* having to go through the formal procedures required to call a court martial.[67] Morison suggested that Hill make official already existing practices which allowed the police to inflict 'summarily and promptly moderate corporal punishment' for petty offences committed by camp followers in the field.[68] Morison felt that this power would curtail the 'disorderly habits' prevalent among the camp followers. It is clear from Morison's statement that, unofficially, commanding officers exercised similar powers and maintained near total penal control over the cantonments. In this way, the perceived needs of the military further shaped notions of what constituted a punishable offence.

Regulation XX laid the ground for such all-permissive control. In addition to making the Articles of War more widely applicable, it required all persons who were resident, or who owned shops operating within the limits of a cantonment, garrison or station to register on a list maintained by the bazaar police. By the order, houses and shops were subject to possible military or police inspection at all reasonable hours.[69] In effect, this formalised a cantonment-wide (and indeed beyond) surveillance operation on ordinary residents and shopkeepers. As military control over Indian civilians spread, the army, in effect, enlisted (or, rather *conscripted*) ordinary individuals, sweeping them into this militarised state.

Ostensibly, such surveillance aimed to curtail the illicit production and trade in liquor and spirits as well as prevent unauthorised 'vice' houses, be they brothels or toddy shops. It required all those wishing to carry on their business within cantonment limits to register. While the Regulation explicitly noted that commanding officers were not authorised to dispossess rightful landowners of house or land in the military bazaar, it gave them the right to make and impose any general regulations on government-owned bazaar land. This series of rights related to the terms or occupation of houses, shops or other 'fixed places.'[70] Following the enactment of Regulation XX, officials generated countless lists relating to cantonment space. Not only were the registration details of those residents or shop owners within the cantonment recorded, but broader lists generated of the cantonment boundaries themselves. The number of official cantonments grew rapidly in the next 25 years. In 1811, the number of officially demarcated cantonments stood at 18; by 1835, this had jumped to 88.[71] Not only were officials given greater control over bodies, buildings and land within cantonments, but the physical space which the military occupied across India was itself rapidly expanding.

With each new regulation, the remit of military power over individuals and space expanded further; the colonial state was content to let this continue. Regulation VII of 1832 in Madras targeted the excessive use of liquor and drugs by European troops.[72] This regulation echoed the earlier Bengal orders, however, it went into more explicit detail about those individuals it was now deemed acceptable forcibly to *exclude* from military areas. Section VI of the Regulation allowed the bazaar police (under the control of the Senior Commissariat Officer of the station) to summon any person within the given area, charged with 'abusive language, slight trespass, inconsiderable affrays ... using false weights or measures, or breach of local regulations'.[73] Such summons would require the accused party to appear at a designated time and place (most likely the bazaar police station) to explain their conduct or receive a warning or reprimand. The unwritten suggestion here is that following an initial summons, the bazaar police or senior officer had the right to expel any unwanted 'troublemaker' from the designated cantonment lines should they continue their objectionable behaviour. Persons of 'notoriously bad character' and those 'suspicious persons without ostensible means of honest livelihood' were by this order also liable to apprehension and expulsion by the bazaar police.[74] Similarly, the officer commanding the station now had the power to remove physical 'nuisances' from military areas.[75] Such nuisances consisted of any buildings or *pandals*[76] found to encroach upon the streets of either bazaar or *pettah*[77] within military jurisdiction.

In order to prevent the smuggling of alcohol and drugs into and around the cantonment, Regulation VII expressly marked for punishment any Indian who attempted to sell liquors or drugs within the limits of the cantonment to any European, unless specifically licensed to do so by the government. The punishment for anyone convicted under this clause was imprisonment with hard labour for up to one month, as well as the possibility of up to 50 lashes for aggravated offences. Perhaps more crucially for the government's coffers, the exclusive control of the sale of liquor and drugs within and around the cantonment was deemed to rest with the Commissary General. Any individual who wished to possess anything more than a bottle of liquor or a seer of drugs[78] now applied to the Officer in charge of the station, who was responsible for granting permits. This officer also held the exclusive power to grant licenses for the sale of liquor and drugs to vendors in and around the cantonment.

Liquor contracts were expensive. One liquor seller in Bombay, Luxemon Mahodjee Roomdar, purchased a contract for the camp bazaar at Suvarnadurg in 1825 at the cost of Rs 759 for a six month period. Contracts required vendors to adhere to a long list of regulations which sought to further 'sanitise' military space. By signing the contract, shopkeepers, in effect, agreed to act as police in their shops. Any 'disorderly conduct' or gaming within the shop was forbidden, and vendors were required not only to prevent such activities but to pass on any information that they might have about 'thieves or riotous persons' directly to both the chowdry and the officer in charge of the bazaar.[79] This suggests that surveillance networks were firmly established in the bazaar – by this Section, the shopkeepers, already themselves subjected to surveillance, became informants. In addition to the above requirements, Roomdar was required to agree to everything from the shop's trading hours to limits on the daily sale allowed to each European.[80]

In 1836, a General Order in Bengal further reinforced the spatial controls available to the military.[81] By this order, anyone hoping to build upon unoccupied land within the cantonment had to first gain approval for the plan from the Commanding Officer of the station. Further, the order noted,

> As the health and comfort of the troops are paramount considerations, *to which all other must give way*, the Commanding Officer will be held responsible that no ground is occupied in any way calculated to be injurious to either, or to the appearance of the cantonment ...[82]

This Order allowed house owners within cantonment limits to rent them to military officers, but imposed strict regulations on the transfer of any property or land. The land, it was clearly stated, belonged *in every case* to the Government and in no instance could individuals sell it.

A further threat to the moral order of the cantonment came from the occasional presence of European or American vagrants.[83] Again, not only was their presence seen to challenge the projected image of European moral superiority in the eyes of respectable Indians, but in some instances, they were seen to represent a physical danger. A nameless European vagrant, assumed to be a deserter from the army, who wandered into the Ahmednagar cantonment and could 'give no account of himself' sparked a debate between the Magistrate, Adjutant General and the Brigadier Commanding the station. In the view of the Commanding Officer, the man was causing mischief within the cantonment, having wandered in, 'lured' a soldier out of the cantonment limits and pressured him to drink (one so 'pernicious' as to endanger his life). He then ambled into both the regimental canteen and the house of the Assistant Surgeon while drunk and launched into tirades of verbal and physical abuse. The Brigadier Commanding at Ahmednagar and the

Adjutant General made clear their willingness to utilise Regulation XII of 1847, which allowed for the arrest and conveyance of any Europeans or Americans found without passports or licenses to the Presidency capital. Accordingly, in order to prevent the man from causing further disturbances in the surrounding areas and in the cantonment itself, he was ordered either to be sent to the Magistrate at Poona or directly to the Presidency under military or police escort.[84]

This series of regulations played an extremely significant role in constructing the idea and physical reality of military space in the early nineteenth century. The increasing boldness of the Regulations suggest that as the Company grew more confident in their control and expansion in India, it was more willing to directly involve itself in the daily lives of a growing number of Indian subjects. In every instance, the reasons given for the introduction of each series of regulations was a greater concern for the health of the European soldier. As exemplified in the 1836 Bengal General Orders, the men's health was central to the decision-making of the military and medical boards. As the European soldiery represented such a costly and essential asset to the Company and Crown, the boards confidently and aggressively laid out their demands regarding spatial and bodily control – for the men's health – in and around the cantonments. This spreading control in turn reflected the militarisation of the colonial state.

In addition to these broader, more formal cantonment regulations, the military also experimented with 'softer' forms of disciplinary control. Mostly, these focused on the production of 'safer' drinks for the soldiery, but increasingly included provisions for alternative forms of entertainment. However, the latter of these were brought forward only in a piecemeal fashion. Until the 1860s, their introduction was dependent on the whim or inclination of individual commanding officers. In one such experiment with the former, however, the Court of Directors sent 15 bags of hops to Bombay in a plan to brew beer for the troops. A committee was appointed to test the results. They pronounced three of the five varieties produced 'good' and 'wholesome'.[85] Similarly, in Bengal it was thought best to supply beer to European soldiers in the hope that this would discourage them from drinking the more potent spirits to be found in the bazaar.[86] The Court of Directors remarked favourably upon another experiment in 1840 to supply the troops of HM's 26th Regiment at Fort William with beer, noting to the Military Board that the picture of the health of the regiment was so favourable that it led them to believe that the whole '... internal economy of the Corps must be in a very perfect state.'[87] They had come to this conclusion by comparing hospital admissions to the number of gallons of spirits sold in the canteen – in this case being some 8,242 gallons less than the government allowance in 1838.[88] This healthy state was deemed to be the result of a well-run hospital, the substitution of beer instead of spirits in the canteen and the establishment of a temperance society. Of course, the reduced level of spirits sold in the canteen did not preclude the possibility that the men were purloining alcohol elsewhere, but the Court of Directors did not explore this eventuality.

Hugh Macpherson, a surgeon in the Bengal army, also observed this supposed down-turn in drink. Macpherson similarly attributed the decrease in drink-related admissions in the early 1850s to the more carefully supervised introduction of beer into the canteen system. He tried to prove this by using statistics from the Bengal Army that tracked the numbers of men admitted for delirium tremens and brain fever in the eight years from 1846. These figures showed a decrease from a peak in 1846 of 8 per cent of overall strength to 3.4 per cent in 1853.[89]

Singing similar praises, the author of 'The European Soldier in India' in 1858 happily noted that a new canteen recently opened in Calcutta on the plain between the fort and the town, so that the men would not be forced to go into the bazaars to seek refreshment. In this way, medical concerns led to racial segregation as men were encouraged to mix only with 'their own' in the safety of the military canteen. To prevent the dangers which so often accompanied intemperance and in common with the aims of the various beer experiments, this canteen provided 'sound, drinkable liquor' as well as a host of other amusements, including quoits, in order to draw the men in for the evening.[90] While no doubt such 'amusements' and readily available beer and liquor did induce more balanced behaviour in some men, it remained to be seen how canteen liquor, priced as it was so significantly higher than that available in the bazaar, could compete.

F.S. Arnott, a Surgeon serving in the 1st European Bombay Regiment in 1854, thought an improvement in the health of the soldiers in the 1850s was the result of a combination of factors. However, Arnott's factors, unlike those of the government, focused on changes to the soldiers themselves. These, in his view, included the recruitment of a higher class of men into the service, better treatment of the soldiers, and the on-going reorganisation and professionalisation of both medical and military services in India.[91] The men now recruited into the Company's service, he asserted, were '... men of great ability and intelligence, of respectable birth, of superior social position, and of excellent education', with this, he noted that, 'sobriety is gradually taking the place of drunkenness and sickness and mortality are yearly diminishing.'[92] Here again, the emphasis rested on the perceived class of the men. Arnott's stress lay on the recruitment of a different sort of individual than had previously been recruited for service in India. He proudly noted (although his claims seem unlikely and are impossible to verify) that every European regiment of the Company's service was provided with a bank, school, library, printing press, coffee shop and 'an excellent theatre'.[93] In addition, other 'amusements' were provided which included chess and cricket clubs, skittles, quoits and draughts-boards. He happily linked the rise in men's deposits in the savings bank with the decrease in the consumption of spirits in the canteen. Arnott and the chaplains were in the minority, however, as most observers on the soldiery actively reinforced the belief in their irredeemably degraded state. Moreover, for every claim that standards (in this respect) were improving, it is easy to find a counter-claim. This we find in an 1857 letter from one British army soldier who assured his brother that there was no place to go in the evening apart from the canteen and coffee room.[94]

Activities to promote the intellectual improvement of the men were similarly limited. The Company began to open libraries for the men across the presidencies in 1819, but it is difficult to gauge their success. They were initially portrayed as a great 'privilege'.[95] It is clear that their defenders – men like Arnott and the regimental chaplains, sought to introduce a higher 'morality' to the men's daily routine. This is reflected in the books permitted as well as in the initial orders, which stated that the books could only be read under the supervision of a librarian.[96] Regimental savings banks were introduced on a broader scale only after 1860 and were similarly restrictive in their policies (and, like libraries, presented as a great dispensation). Soldiers were required to give at least seven days notice to their Commanding Officer if they sought a withdrawal of any or all of their savings. Further, if that officer believed that the individual intended to make 'improper' use of his money, he was authorised to withhold this 'privilege' of withdrawal.[97]

Another alternative was proffered by the numerous temperance societies that sprang up in cantonments and towns from the late 1830s. These groups were usually promoted by regimental clergymen and missionaries and required their members to take a pledge, whereby they would promise to abstain from drink and contribute the money they would have spent on this 'vice' to social or religious causes. The societies' success in curbing intemperance varied. Often, as was the case with the men of the 18th Regiment in Bengal, temperance and 'virtue' levels were inconsistent over time. In this Regiment, a number of men signed a temperance pledge while in Fort William but, on learning that they were to be despatched to Burma, begged the clergyman who had administered the pledge to be relieved from it. This plea was made on the grounds that they felt they would not be able to tolerate the climate of Burma without the assistance of liquor.[98] Temperance societies within cantonments were ordered abolished by the government in 1845, on the grounds that clubs and societies were not permitted within the army, and their presence violated this rule. The societies remained in towns (and their strength grew markedly in the latter part of the century) but, from 1845, they could no longer be as close to the soldiery as they might have hoped. No material remains that could outline the reasons behind this decision. However, concurrent debates suggest that the army was not interested in wholly stopping the men from drinking, as this was seen to be an integral part of the soldier's composition. Like Wellesley's disdain for educating soldiers, in the eyes of many commanding officers, temperance societies' attempts to reform and uplift the soldiery threatened to undermine the men's brute strength.

The military made further alterations to encourage temperance through such measures as good conduct pay, which was first introduced in 1837. Company troops were entitled to an extra penny per day after seven years of service, an amount that increased again after 14 and 21 years.[99] This measure was extended to Crown troops in 1838. However, this system of conduct rewards was not without its problems. Until the Crown takeover in 1858, there remained some disparity in the levels of rewards administered to men in the Crown and Company armies and indeed, as earlier *batta* protests show, across the three Presidency armies. While invalids in the 1850s in Her Majesty's army could expect to receive an increase of pension based on good conduct, those men serving in the Company's army failed to receive this bonus.[100] Naturally, and like other pay-related issues such as *batta,* this provoked some resentment among the men.

In 1858, the military released new sets of guidelines for sanitation and the soldier's 'mode of living'. These were more basic and laid out *suggestions* on lifestyle choices such as the advisable times at which to sleep and the ideal thickness of clothing.[101] These guidelines warned soldiers that, 'A drunkard, or even one who drinks spirits freely without being intoxicated dies early in India.'[102] The military printed these sanitary commandments in both pamphlet and poster form and placed them in the barracks for the men to read. However, concrete physical plans, such as those proposed by Clark, Ainslie and Chevers remained unrealised.

Disorderly European women

For those attempting to forge a new medical and moral order within the cantonment, it was all too easy to blame the supposedly untrustworthy Indians 'lurking' in bazaars whose 'natural habits', it was alleged, predisposed them to vice. However, when the guilty party was a European woman, blame was quickly reconstructed to suit a separate, but complementary set of class-based assumptions. Some of the European wives and widows of soldiers sold liquor illegally in the bazaar or ran small operations out of their quarters. It was alleged that a number also prostituted themselves, at times with the understanding or support of their husbands. These facts raised a complex set of problems for commanding officers and the government alike. Such violations came at a time when there were still active debates surrounding the wisdom of allowing increased numbers of European women into India. As discussed in earlier chapters, the presence of European women in India raised a number of fears for many Company officials. Many felt that such an influx had the potential to severely disrupt the recently attained balance in society. In addition, European women were generally believed to be too 'delicate' to withstand the rigours of India and therefore the accompanying unavoidable medical costs concerned the Company. Both debates shared a common vision of the European woman. The image of the 'pure' and fragile European female was used by both early Company officials in order to prevent their entry to India as well as by later imperial propagandists who encouraged their 'domestication' of the empire.[103] As Ann Stoler has argued (with regard to Southeast Asia) the attitude towards 'poor whites' and white women were closely linked and inherently problematic.[104] Similarly, in India, problems swiftly arose when the very same European women were accused of a wealth of vices within the Indian cantonments. It was imperative therefore, that such women were portrayed as symbolically (and often physically) separate from the desired ideal of European womanhood. In a strikingly similar manner to how Indian women deemed to be 'prostitutes' were dealt with, 'dissolute' European women were systematically ring-fenced and pronounced to be both of low-class and bad character and were often ordered to be physically removed from the cantonment. It was stressed that such women were anomalies, distinct from the 'respectable' and 'quiet' women of the regiment.

Opinion as to the reasons why some European wives of soldiers carried out activities such as the illegal sale of liquor and prostitution in cantonments remained divided. On the one hand, some observers suggested that the 'type' of European woman who came out to India to find a husband was of the same degraded moral stature as the men themselves, and accordingly, she 'naturally' engaged in these activities. On the other hand, was a smaller group who argued that the pay of the married soldier was so miserly that the women were forced into such activities to supplement their husband's earnings and support their families. An 1825 correspondence between GA Weatherall, Secretary to the Commander in Chief at Fort Saint George and Thomas Munro, Governor of Madras, captures these two arguments. On writing his proposals for checking the illegitimate sale of liquor within the cantonment, Weatherall argued that the illicit sale of liquor within cantonments would quickly be checked if the government were willing to grant an increased allowance to the wives and children of soldiers.[105] Munro responded by swiftly dismissing Weatherall's claim, noting, 'If the selling of spirits were confined to women having families the argument would be conclusive, but ... I believe that the practice is owing less to real want than to abandonment and the thirst of gain.'[106] Clearly belonging to the military school of thought which sought to restrict the number of European women entering India, Munro appeared to maintain a vision of a military space denuded of women as essential to maintaining Company rule in India. It is no coincidence that in this same Minute, he also dismissed requests to provide greater support to the half-caste families of European soldiers. Such relief, Munro claimed, would only encourage a proliferation in their already large numbers.[107]

Munro's obvious prioritisation of economy and restrained spending pointed to the economic concerns present throughout the 1820s and 1830s. However, his response highlights just how important economic considerations were in determining approaches to medical, moral and social control within military spaces. Munro suggested that his decision had been made on 'moral' and 'humanitarian' grounds, however, his preoccupation with cost reveals that such decisions were never clear-cut. Military and government decision-making operated in a space restricted by a number of important parameters, including fiscal and health concerns. Munro's suggestion for a solution to the problem of soldiers' wives selling liquor was simple – restrict the number of European women allowed to come out to India, as both the cost of maintaining such a body of European wives and children had become a heavy burden to the Company.[108]

The anonymous author of 'Soldiers; their Morality and Mortality' further corroborated Munro's vision of the dissolute, greedy and lustful soldier's wife:

> There are two classes of soldiers' wives, – those who have been born in India, and those who have been introduced from Europe. The

> former, particularly if their skin is of a darker tinge, are rather looked down upon in Her Majesty's regiments, but we are bound to say that this is most unjust. They are usually superior, intellectually and morally, to their European sisters... The wives who are brought from England have been for the most part inhabitants of sea-ports and garrisoned towns, but a few were once rural maidens, ignorant of the world until they saw one of its phases in a barrack.[109]

He goes on to insist that many of the women, when faced with intemperate husbands, simply fell in with the common habits in the barracks and in this way, quickly transformed into 'Doll Tearsheets'.[110] The only women of the regiment who managed to escape this fate were those who refused to mix with the other women within the regiment.[111] In these instances, the 'low-class' European woman was seen to be 'naturally' inclined to such activities. Yet, pivotally, unlike her husband, these wives were deemed eligible for punishment for such activities.

Regardless of where an observer's sympathies lay, European women who were repeatedly caught selling liquor (or engaging in illicit prostitution) were seen as hugely disruptive to the good moral order of the cantonment. To solve this dilemma, commanding officers often requested permission to expel such women from cantonment lines. Lone women who had left their husbands and the widows of soldiers were as threatening as the single daughters of soldiers, all of whom could easily disrupt the equilibrium of the cantonment. Lieutenant Colonel Henry Havelock, in 1848 serving as Deputy Adjutant General of Bombay, wrote of one such woman in the Deesa cantonment, whose husband was serving in Karachi and whose presence in camp was, in his view, problematic.[112] Havelock argued that it was within the rights of the Adjutant General to forcibly expel her from the cantonment, but worried that if she roamed freely about the surrounding areas, her dissolute behaviour would serve to 'disgrace the European character in the eyes of the natives'.[113]

The practice of expelling 'dissolute' European women from the cantonments and sending them to Europe was a well-established one. General Orders issued in Madras in 1813 reveal that such women had, until that point, been treated as prisoners until their return to England could be arranged. The 1813 orders discontinued the practice of imprisoning the women at Poonamallee, but maintained the government's right to forcibly return them to England.[114] The objection made to detaining the women prior to their departure mainly came on account of expense. However, the practice of enforced return remained a favoured option and women deemed to be 'dissolute' or 'irregular' continued to be handed over to the custody of the Town Major to await their deportation to England.

The idea of a more punitive control over 'unprotected' European women was highly appealing to many military officials. Such women (either loosely married or widowed), who failed to comply with the moral imperatives dictating married life in the cantonments, were often accused of 'idleness' even if they were not themselves involved in illegal activities. In 1822, the Madras Superintendent of Police went as far as to suggest that all women, not 'belonging' to the troops be required to register with the Town Major and be subject to surveillance.[115] In this way, any improper behaviour could be swiftly detected and it would then be easier to report women whose behaviour fell foul of cantonment rules. The Superintendent's plan extended beyond 'unattached' women, however. He suggested the issuance of Regimental Passports to all European wives and widows of soldiers and officers, which they would then be required to produce when travelling outside regimental lines.[116] However, even without such official sanction, the cantonment police often kept single and widowed European women under close, unofficial surveillance. When asked to produce a detailed list of 'troublesome' European women for the Secretary to Government, Mr Ormsby, Superintendent of Police at Fort St George in 1823, quickly produced a detailed list of the women's whereabouts and activities in reply (see Table 4.1).

Upon receiving Ormsby's list, the Secretary to Government ordered a sea passage back to England arranged for the six named women. However, the women themselves, who appeared to have little desire to return to England, foiled these plans. Upon receiving written notification from Ormsby that they were to prepare for embarkation, Eliza Williams and Mary Ann McMullen quickly wrote back stating their inability to return due to their health (Macmullen was pregnant while Williams noted her condition to be 'delicate').[117] Ms Burns wrote Ormsby a similar letter from Wallajahabad, sending her regrets, but noting that she had left Fort St George on account of her ill health for a much-needed change of air. Mrs Pye, like Mrs Thompson, was found to be convalescing in the General Hospital. No specific report was given on Mrs Cunningham, although from Ormsby's remarks, she appears to have been the most 'threatening' of the six, as she was found to be directly engaged in prostitution and begging. Similarly, this surveillance and deportation order was in evidence in Poona in 1824 when two women who were alleged to be committing 'the greatest irregularities in their huts long after hours' were promptly ordered deported.[118] In

Table 4.1 European women who lead a dissolute life in Madras, as also of those left without protection

Names	Whether married or not	To what regiment belonged	Ostensible employment	Remarks
W Burns	Unmarried	The Royals		The leader of a dissolute life
Eliza Williams	Unmarried	From New South Wales formerly kept by Ensign Ross of HM 46th Regiment		Ensign Ross having left her, she has since lived with another Gentleman by whom she has been left with two children and has now no protector. The dates of the correspondence with Government regarding this woman are 6 October and 4 December 1818. 18 & 19 January, 4 & 16 March 1819
Mrs Stone alias Pye	Unmarried	HM 25th Light Division		She is kept at present by Sergeant Pye employed at the Deputy Adjutant General's Office King's Troops
Mary Ann McMullen	Unmarried	HM 80th Reg't		This woman was with Mr W Parr before and he has undertaken to send her home
Mrs Cuningham	Married	The Royals	None	Her husband is at Trichinopoly with his Regiment and she has been at Madras for some years prostituting herself, and sometimes begging, etc.
Mrs Thompson	Married	The Royals		Her husband is in England, and she is reported to be now in the General Hospital.

Source: *Letter to the Secretary to Government from Mr Ormsby, Superintendent of Police, 22 February 1823*. TNSA Fort St George Public Consultations, 28 February 1823.

other instances, where the Town Major had arranged transportation for women back to England, they fled before their ship was due to sail.[119]

Women who remained married to their husbands, but who displayed signs of 'abandonment' posed an additional problem. In such cases, the military sought to protect the men. Men serving in higher positions were seen to have their reputations at stake, but even in disputes between soldiers and their wives, the decision on the woman's right to remain in India remained firmly with the husband. 'Abandoned' wives were routinely sent back to England after first obtaining the permission of their husbands.[120] Husbands, having fallen out with their wives, could also request their return to England, citing the women's behaviour or general misconduct. Should a wife remain in camp without permission while the regiment moved on and later wished to join her husband, it was entirely left up to him as to whether she be allowed to do so. If he refused, government sought his permission for her removal to England. It would appear that most local magistrates willingly agreed to this compulsion. However, there were a handful of dissenting opinions. In 1865, remarking on the case of a Mrs Philips, the wife of a sergeant in the army, the Judge Advocate General wrote his opinion that neither civil nor military authority had the power to deport her by means of compulsion or without her consent. Nevertheless, in response to this, the Officiating Secretary to Government noted that although it was not permissible to use violence to compel Mrs Philips to leave, her husband's consent would serve as a bar to any civil action taken against the Government.[121] In a separate case a few years later involving two women in Ahmedabad who were reported to be bad characters (by virtue of their heavy drinking, violent tempers and obscene language), the Adjutant General ruled that with the consent of their husbands (one of whom was serving in Abyssinia and the other in the Bombay jail), they could be sent to England.[122] This line was, however, blurred somewhat when the question of the daughters of soldiers (many of whom had never lived in England) were accused of leading 'dissolute' lives. This once again raised the thorny question of what it meant to be a 'European'. Here, it became more difficult to force women 'home' who had never set foot on British soil.

In 1862, the Government returned to the idea of a designated prison for dissolute European women. One proposal questioned if it would be more efficient to establish such a jail for European women in the hills, or at least in a 'healthy' location far enough away from any military stations.[123] However, in response to this query, the Quarter Master General at Fort St George returned to the ever-important concern of cost. He felt that the cost of sending the women to England was still cheaper than establishing and maintaining a separate jail for them, and, as such, the idea appears to have been set aside. A similar bill was proposed and rejected in Bombay in 1865, again on the grounds of expense.[124]

European women living in the cantonments in India in the early nineteenth century occupied a curious position both socially and legally. As a symbol of white womanhood, they were held up as supposed examples of civilisation, yet the reality in many cantonments was much more complex. As in England, many women moonlighted in more marginal occupations in order to supplement the meagre incomes of their soldier husbands. Army and government alike resented these informal sidelines and deemed them illegal. As such, the army blamed these European women for disrupting the order of the cantonment and endangering the men's health. Punishment for carrying out illegal activities in the cantonment which led to the men's intemperance divided along racial lines. Indian camp followers and cantonment dwellers were subject to surveillance, corporal punishment and expulsion from the camp for any 'disorderly' behaviour. While European women living in the camps were still subject to surveillance, there is no evidence that they were ever subject to corporal punishment (nor was the possibility ever entertained). The preferred method to discipline these women was the threat of expulsion from the camp and forced return to Britain. Although, both punishments were unquestionably unpleasant, the fact of a difference between the two further reinforced the idea of racial difference.

Courts martial and punishment

The exponentially increasing punitive power over Indians and European women handed to military officials was, ostensibly, to protect the health of the soldiers. However, this period did not witness a similar expansion in their power to punish the European soldier for placing his own health at risk through intemperate behaviour. This begs the question – which offences were solders formally punished for? When army and medical men lay so much stress on health and physical threats from without, what was their response when the men brought harm upon themselves or others through their 'misbehaviour'? Punishment of soldiers in the cantonment was administered in a number of ways, from the informal (for example, assigning extra duties or withholding pay), to the formal (the Regimental, District or General Courts Martial). However, with drink being such a concern for officers, it is important to first examine how the army dealt with drunkenness and misbehaviour.

As suggested above, anxiety about the level of drunkenness amongst the troops was not strictly limited to medical concerns. The predominant belief among external observers and officers alike was that drink fuelled the majority of crime committed by soldiers. It would appear that such fears were justified. Observers on the subject of drink argued that the majority of courts martial on European troops were on account of the men's habitual drunkenness. A conduct report on the 1st European Light Infantry noted that although the regiment was generally well-behaved, a spate of courts martial took place in the four months following the regiment's return from Afghanistan. During this time, the report recounted, the men, having received their back pay, drank more than normal and an outbreak of crime ensued.[125] While this swell of courts martial is impressive in itself, what is perhaps even more significant is the fact that most offences and incidences of 'bad behaviour' and drunkenness within the cantonment were punished *without* recourse to court martial.[126] The 'character books' kept in every regiment on the men support this fact. In these books, officers carefully recorded any misconduct by the men along with accompanying punishments.[127] These books reveal that a badly behaved soldier often needed to commit a number of offences before any formal action was taken to castigate him. Monier Williams and the author of 'Soldiers; their Morality and Mortality' both cited examples of drunkenness and punishment to support this claim that most punishments were the result of drink (see Table 4.2).[128]

Unfortunately, because few accounts survive, it is difficult to more broadly confirm these assertions. From the few records which are available, however, it would appear that there is strong evidence to support these claims. More interesting is the fact that those men accused of drink-related offences were not always convicted by courts martial. Acquittals often resulted even in seemingly clear-cut cases (such as when men repeatedly appeared drunk on duty or suffered from delirium tremens).[129] However, this fact seemed to depend upon a number of factors. These included (but were not limited to), the rank of the offender, whether the bouts of drunkenness were conducted in a public or private place (i.e. in town or on regimental parade versus in the barracks or canteen) and whether any additional misbehaviour or violence was also involved.

A clear distinction existed between attitudes towards those soldiers court martialled for drink or bad behaviour and officers accused of the same. For many of the cases in which soldiers were called before a court martial, the case had been brought only following especially severe misconduct and/or repeated bouts of public drunkenness. With higher officials, on the other hand, the public nature of their bad behaviour was a far more damning offence. Intemperate behaviour from a man in

Table 4.2 Drunkenness and punishment in Madras, 1847

Corps	Strength	Number of drunkards	[percentage] of drunkards	Number punished by Regimental Captains and Commanding Officers	[percentage] of soldiers thus punished	Number tried by Regimental, District and General Courts Martial	[percentage] of soldiers thus tried	Total of all classes of offences	[percentage] of offences
Artillery	1,958	345	17.62	1,318	67.31	162	8.27	1,480	75.58
HMs 15th Regt.	658	45	6.84	265	40.27	11	1.67	276	41.94
HMs 25th Regt.	716	88	12.29	615	85.89	38	5.31	653	91.20
HMs 51st Regt.	1,089	203	18.64	507	46.55	22	2.02	529	48.57
HMs 84th Regt.	1,062	136	12.80	731	68.83	19	1.79	750	70.62
HMs 94th Regt.	1,137	315	27.70	659	57.96	55	4.84	714	62.80
HCo's 1st Eur	1,048	155	14.79	578	55.15	36	3.43	614	58.58
HCo's 2nd Eur	1,075	174	16.18	1,431	133.11	92	8.56	1,523	141.67
Total	8,743	1,461	16.71	6,104	69.81	435	4.97	6,539	74.78

Source: 'Soldiers, their Morality and Mortality,' 196.

a position of power suggested a loss of order and related to this, a threat to public confidence in the moral and physical ability of Europeans to control India. Bad behaviour, it was implied, would place Europeans on a similar moral level as Indians; this was to be avoided at all costs. However, this is not to imply that soldiers were dealt with any less severely for their 'crimes', but simply that such transgressions were more often punished using informal means. Legal observers believed that the public shame which accompanied a court martial was much more effective in disciplining men of officer rank. Again, this highlights the 'moral' and class separation of officers from the rank-and-file. An officer was seen to be an individual, a man of dignity and honour. Such a man would naturally be ashamed by the process of a court martial. The soldier, on the other hand, was thought to be much more as a slave to his natural (read: carnal) impulses. In this light, as a more animalistic creature, a process which called his dignity into question would less easily abash the soldier. Therefore, in common with Wellesley's argument about corporal punishment, most military officials saw more tactile, physical punishments as more effective for the rank-and-file.

A public display of 'insensibility' was enough to strip more senior officers of their positions and earn them a discharge them from the service. Such was the case for Lieutenant Dick of the 74th Regiment Native Infantry who was charged with 'conduct scandalous and highly disgraceful to an officer and a gentleman' in having been drunk on multiple occasions and, in consequence, repeatedly taken in by the Calcutta town police.[130] There is no suggestion that Lieutenant Dick behaved aggressively or committed a physical offence, per se. The core of his crime was in the public nature of his drunkenness – in the shame of an officer being seen inebriated on a street in Calcutta. Such a display, which was expected of the 'lower' ranks, was simply unacceptable for men who had reached the rank of a lieutenant (and, perhaps more importantly in Dick's case, one assigned charge of the native troops).

It was not just officers who were held to a different standard than the rank-and-file. This divide regarded surgeons as part of the officer class. The punishments handed down to surgeons for misconduct reinforces this point. In addition to being charged with conduct 'unbecoming' and 'disgraceful', Assistant Surgeon Alexander Storm of the 51st Regiment, Native Infantry, was charged with a raft of other offences, including contempt of authority, disobedience of orders and neglect of duty.[131] From what remains of the proceedings of his trial, it appears that Storm had appeared drunk in front of the Commanding Officer. When accused of the offence the following day, he had responded rudely and violently. Following his arrest (which he resisted), Storm barged in, drunk, to the quarters of the brigadier commanding his regiment. Finding Storm guilty and sentencing him to suspension from rank, pay and allowances for six months, the court noted that they had issued a 'lenient' sentence. They made clear that they had granted such clemency on the grounds that Storm was a habitual opium-user (purportedly for medicinal purposes) and they considered it most likely that his state of intoxication was concomitant with the effects of the drug.[132]

While the military courts martial severely sanctioned officers for their intemperance, a measure of bad behaviour appeared to be expected of the soldier. Courts martial against the rank-and-file often involved a repeated pattern of bad behaviour or the commission of a serious offence. Typically, punishment for general misconduct was by less formal means, including extra duties or pay stoppages. The character book of one Gunner, William Carter, of the 1st Company 1st Battalion Artillery, provides particular insights in its illustration of different types of punishment inflicted (see Table 4.3).

Table 4.3 Extract from the Character Book of the 1st Company, 1st Battalion Artillery, Agra, 28 February 1845

Gunner William Carter, No 2934. Enlisted 23 September 1841, Landed in India 4 March 1842. Joined the Company from 2nd Battalion Artillery, 17 August 1842

Occurrences and wounds when and where received	Remarks and punishments awarded
1842	
25 March. For being from Barracks, from 10 o'clock pm and not returning until 9 pm 26th instant	Six days fatigue drill, on heavy marching order
7 April. For direct disobedience of orders in quitting the barracks after hours, and attempting to pass the boundary guard to Calcutta, with a false chit between 11 and 12 pm 7th instant	Seven days marching drill and two extra guards
18 April. Absent from drill pm, 18th instant and not returning till between the hours of 5 & 6 pm 21st instant. 2nd for being deficient of two shirts and one pair of pantaloons of his regimental white clothing	To have no pay for the three days he was absent and to be sent to the Congee House* for seven days.
21 May. Absent from his cot between the hours of 10 and 11 roll, and not returning till between 1 and 2 am the following morning	Seven days Congee House
26 June. For disgraceful conduct in having at Dum Dum, on 28 June 1842, stolen a regimental undress** jacket, the property of Francis Hedley, Gunner of the 5th Company same battalion and regiment	Tried by District Court Martial on the 21 July and acquitted
13 August. For absenting himself without leave from his detachment, on or about the 8 August, and not returning until brought back by a guard from Fort William on 13 August 1842	By a Regimental Court Martial, to be imprisoned for 20 days, and to have five days pay stopped
23 November. Drunk, when for instruction drill between the hours of 4 and 5 pm 23rd instant.	Seven days Congee House
16 December. Absent from instruction drill, at 4 o'clock pm, 15th instant, and in not returning until between the hours of 5 & 6 am on the morning of the 16th instant	Seven days Congee House and three days drill

(*continued*)

Table 4.3 Continued

Gunner William Carter, No 2934. Enlisted 23 September 1841, Landed in India 4 March 1842. Joined the Company from 2nd Battalion Artillery, 17 August 1842	
Occurrences and wounds when and where received	**Remarks and punishments awarded**
27 December. Absent from gunfire roll call am 27th instant, and not returning until brought back from Chitpore Road, by an escort of the 3rd battalion between 6 and 7 am same date.	Seven days Congee House and seven days drill.
1843	
27 February. Absent from between the hours of 4 & 5 O clock pm 24th instant, and not returning until being brought back by an escort from the Battery Guard Fort William between the hours of 7 and 8 o clock pm 27th instant	30 days punishment drill, seven days Congee House, and three days pay stopped
1844	
14 August. For being drunk on guard, between the hours of 6 and 7 pm 14th instant	Seven days Congee House and seven days liquor stopped
18 October. For fighting in barracks	Six days defaulters' drill
1845	
27 January. Tried by a District Court Martial for disgraceful conduct in having on the night of the 18th of January in camp at Agra, feloniously opened a box, and stolen there from one bason*** [sic], one plate, and one loaf of bread, the property of 13 mess of the 1st Company 1st Battalion	Found guilty with the exception of the words 'one bason, and one plate', sentenced to imprisonment for three lunar months and to be kept in solitary confinement on the second month, not confirmed; the proceedings being vitiated by the omission of putting the prisoner on his defence.

* A Congee House was a small, solitary confinement cell where the prisoner would be fed on a 'congee', or rice porridge diet.

** In military terms, an 'undress' uniform was worn by soldiers for normal duty, while a 'dress' uniform was those worn for special occasions, or regimental parades.

*** *Hobson-Jobson* defines a 'bassan' as a dinner-plate.

Source: *Extract from the Character Book of the 1st Company, 1st Battalion Artillery, 29 February 1845*. NAI, Home (Military) Proceedings, 28 March 1845, no. 82.

From this, it appears that Carter was in the habit of 'wandering' out of the cantonment, an action frequently lamented by surgeons in their reports due to the high level of illness which often followed such peregrinations (the understanding being that men either wandered off in search of drink, women or other forms of 'dangerous' amusement, and returned drunk, venereal or both). Interestingly, despite his wanderings, Carter was punished only twice for drunken behaviour, first on 23 November 1842, while he was on instruction drill, and second on 14 August 1844, while he was meant to be on guard. Carter's other instances of drunken behaviour were overlooked, but these two instances were targeted as he was, after all, meant to be on duty. As such, these instances of public drunkenness potentially threatened the general discipline of the corps. Carter's numerous absences from barrack or cantonment, while possible court martial offences, instead earned him time in the Congee House or additional drill. Indeed, his liquor allowance at the canteen was stopped only after his second instance of drunkenness. Carter's repeated misconduct did, however, add up; following the theft of 64 rupees from the kit box of a corporal, a District Court Martial sentenced him to corporal punishment and dismissed him from the service with ignominy.[133] The pattern of punishment evident from Carter's Character Book is further supported by the more general record kept of drunkenness and punishment in Madras in 1847 (see Table 4.2). These figures suggest that the majority of drink-related offences were informally punished by captains and commanding officers, with only a relatively small percentage of those (though still significant numbers) progressing as far as the court martial stage.[134] While the number of soldiers brought before a court martial on drink-related offences was significant, as evidenced by the Madras figures, it represented only a very small number of those punished in other ways for intemperance. Nevertheless, despite these numbers, it is important to remember that intemperance was not the only punishable 'moral' crime within the cantonment. The pattern of these other offences helps to cast light upon the ways in which military officials sought to morally order the cantonments through the administration of punishments. When it came to the soldiery, officials picked their fights carefully. Clearly, the fact that individuals like Carter, who committed multiple offences (many of which were grounds for dismissal), were retained suggests that the army was desperately short of suitable European recruits. The quality of the recruits made the Company all the more vulnerable. Not only was punishment adapted for fear of invoking mutiny, but, once again, it exposed the uneasy control which officers exerted over their European soldiers.

Disgraceful and unbecoming conduct

Although disgraceful and unbecoming conduct often involved drink, this charge acted as an umbrella for a range of other offences (usually of men of the higher ranks), which included everything from unpaid debt to sexual misconduct or slander. A General court martial found Brevet Captain George Becher of the 4th Regiment Bengal Native Infantry guilty of slander and conduct unbecoming after he launched a campaign to suggest that his Lieutenant Colonel, Henry Coley, had conducted a relationship of 'unusual intimacy' with one of the boys of the regimental band. Becher, Coley asserted, had not only suggested this relationship to other junior officers, but had written 'anonymous' letters to the *Delhi Gazette* and *Agra Ukbhar* newspapers posing the same accusation. For this, Becher was found guilty and cashiered,[135] but not before the court lamented that such a 'heinous' accusation had been connected with the good name of Colonel Coley.[136] Perhaps unsurprisingly, cases involving accusations of 'unusual' or 'unnatural' sexual behaviour often revolved around contested or resented power dynamics within the regiment. This could range from the high-level cases such as that of Becher to those which more closely resembled ritualistic indoctrination practices. Although not explicitly mentioned in many army accounts, incidences of bullying were likely just as common among new recruits in India as they were in public schools in England. Most cases of abuse appear to have gone unpunished, but a handful of the more extreme were elevated to court martial level. Of these, sexual impropriety (or the threat of impropriety), accompanied by violence, helped to ring-fence the most severe cases. One such case was that of Ensign MacNeill of the 72nd Regiment Native Infantry, who, while proceeding up-river from Calcutta to Dinapore in December and January 1844–1845, severely bullied one of his fellow ensigns, Trenchard.[137] Trenchard alleged that MacNeill beat him with a stick, extorted items from him and forced him to drink (while he had been ill with fever). However, the final proverbial straw came when MacNeill attempted to force Trenchard into acts of bestiality. While his other acts of bullying (threatening as they were) might have otherwise gone unreported, it was the possibility of bestiality that set this tormenting apart. The court found MacNeill guilty and ordered his dismissal from the service.[138] Sexually 'abhorrent' practices carried with them a much more serious charge and punishment than (to a modern observer) seemingly lesser crimes. Sodomy and bestiality were, in the Articles of War, considered together and the punishment for both was death. It is difficult to determine the exact number of men either accused or convicted of 'unnatural offences' as courts martial for such offences most likely took place in closed court sessions.[139] However, even the threat of a possible death sentence for such offences is significant in itself.

To put this punishment into context, 'unnatural' crimes were considered much more serious than rape, which was in contrast punishable by corporal punishment and *possible* dismissal from the service.[140] In addition, rape claims required detailed evidence from the victim which was often difficult or impossible to obtain (including evidence of both 'emission' and penetration).[141] Remarking on the importance of the observations of the local surgeon in rape cases, William Hough, in his 1825 manual on Indian courts martial remarked that rape was a crime, '... of which the accusation is peculiarly easy and the disproof is proportionately difficult'.[142] Hough's assertion and the sentencing guidelines in the Articles of War suggest that the sexual violation of men (as well as consensual sex between men) was considered to be much more serious a crime than the rape of women. In the cases of Becher and MacNeill, even the *suggestion* of the possibility of such an act was enough to merit both men being cashiered. In 1858, a new set of *Rules and Articles for the Better Government of the Officers and Soldiers in the Service of the East India Company* was issued which added further punishments for cases of 'disgraceful conduct'.[143] Article 87 of these rules targeted those men engaged in conduct '... of a cruel, indecent or unnatural kind.'[144] In addition to the punishment handed down (assuming it was less than a death sentence), convicted men were liable to forfeit any claim of pension, good conduct pay and any savings they might have accrued in the regimental bank and were liable to be discharged with ignominy from the service.[145] These 'unnatural' acts were especially threatening as they were the antithesis of vital masculinity. As Mrinalini Sinha has argued, the idea of 'appropriate' masculinity was intimately linked with the construction of power in colonial India.[146] Such a 'void' of masculine behaviour could undermine the military position of the state in India.

In addition to attempting to prevent such 'unnatural' acts, surviving courts martial also inadvertently reveal the many intrigues that took place in the cantonment. Courts martial dealing with 'improper conduct' often featured the wives of European soldiers. The women themselves were rarely tried (unless for outright or attempted murder)[147] but were often mentioned (if not always by name) in cases of infidelity, bodily harm or murder. In a petition to the Governor General, Thomas Pacey, then in Calcutta jail awaiting transportation (for life) following his conviction for murdering a Sergeant with whom his wife had been having an affair, revealed not only the intrigues, but some of the information networks at work within the cantonment.[148] From Pacey's petition, it transpires that (unbeknownst to him) his wife had been conducting an affair with Sergeant Robert French and had run away to live in his bungalow. In searching for her, Pacey sought French's advice, who assured him that his wife had gone to Sialkot. As he was about to set off for Sialkot (after borrowing a pistol and a horse for the journey), one of Pacey's young children intervened and revealed her mother's true whereabouts. The daughter, it appears, had received this information from 'a native', in all likelihood, one of their household staff. Pacey does not mention the source of this information, but it is possible that it came from someone within French's bungalow. That this information might have been common knowledge to their respective staff is not surprising, however, the fact that it was transmitted to Pacey's 'young' (her age is not specified) daughter seems somewhat unusual.

Affairs such as that between Sergeant French and Pacey's wife were problematic for another reason. They broke down the invisible boundaries meant to exist between those of lower and higher rank. Had Sergeant French not succumb to his wounds, he could have just as easily been tried for unbecoming conduct for his part in carrying out the affair. Those in higher ranks, it was held, were not only to set an example for the men, but were to hold positions of trust. One lieutenant, referred to only as 'R.C.' of the 19th Foot at Colombo, was tried for 'degrading himself' by having an affair with the wife of a private of the same regiment.[149] After confirming R.C.'s 'guilty' sentence, one of the Lieutenant Generals conducting the court martial remarked that R.C.'s behaviour was so dangerous and reprehensible as, in carrying out this '...open and avowed criminal intercourse with a soldier's wife,' R.C. had degraded 'himself to a *level* with the soldier, extinguishing all that respect for his officer'.[150]

Conclusion

The soldier seemed destined to maintain his 'lowly' status. From attitudes and measures taken to control intemperance to the punishments themselves, in the first half of the nineteenth century, officials viewed the European soldier in India as little more than a fighting machine. This belief was reinforced time and again by both military and medical observers. Yet, as he represented such a high cost for the Company, control over his health and behaviour was of vital importance. This meant that officials tread a fine line between policies that ensured that brute force and 'masculinity' were maintained while altering the space around the soldier. Measures to expand military control within and beyond the cantonment were frequently justified on the grounds of the men's health.

Central to these measures was the ever-present anxiety surrounding what is best described as the social and leisure activities of the soldiers. Linked, directly as they were, to his health and therefore productive ability, there was an effort made, beginning in the early nineteenth century, to control and manage these activities. The low esteem in which the soldier was held helps to explain why proposals to shift the focus from alcohol to 'healthier' drinks and activities such as gardening, theatres and reading, were not adopted by either the Company or Crown for much of the first half of the nineteenth century.

Yet, as fearful as officials were of the effects of alcoholic and bodily intemperance, there remained a belief that both were somehow inherent to the soldier's very being. This juxtaposition goes some way towards explaining why it was that little real progress (or even effort) was made in reducing the amount of alcohol which soldiers consumed. Instead, it clarifies the alternate approach adopted – to regulate the 'others' in the cantonment who supplied the men with both drink and sex. In this light, the carefully balanced 'proper' functioning of the cantonment was seriously threatened by both those individuals outside its boundaries who attempted to lure soldiers away, as well as those who represented a 'fifth column' within, such as the wayward wives of European soldiers. Those European wives who participated in 'immoral' activities (such as prostitution and the illicit sale of liquor in the barracks) were the focus of extreme anger and censure from both commanding and medical officers. Similarly, arrack vendors in and around the cantonments who operated without government permits not only deprived the treasury of valuable revenues, but were thought to produce a much more lethal variety of spirits.

Perhaps most importantly for the formation and understanding of the empire, however, were the consequences of this approach. As countless invasive regulations were justified on the grounds of protecting the health of the European soldier, the Company pushed out the boundaries of the actual physical space under their control. Medical (and moral) demands made by officers, were quickly fortified and validated through regulations that expanded military space and gave greater, more absolute punitive power to commanding officers not only over soldiers, but all those who came into contact with them.

5
'Unofficial' Responses to Lock Hospital Closure, 1835–1868

Earlier chapters have described how the lock hospital system and cantonment regulations developed in response to a number of fears. Venereal diseases and the women who supposedly transmitted them were seen to threaten the state on many counts. An attack of syphilis effectively rendered the European soldier useless; treatment could take weeks, if not months, and the already budget-conscious Company begrudged the expense involved. Moreover, the behaviour of Indian 'prostitutes' was viewed as a different sort of military and social threat. In a similar manner, those who 'lured' the soldier to drink were seen to disrupt the line of control required to maintain military discipline. If left uncontrolled, it was feared that both would negatively influence the discipline of the troops and further, scandalise 'respectable' Indian society. Supporters of the regulatory system successfully argued that in times of military and political uncertainty, careful regulation was essential to minimise the risks faced by the army.

The political climate had shifted somewhat by the 1830s. The aggressive military campaigns waged in the early nineteenth century had abated. The wars in Burma, Afghanistan and the Punjab which occupied Company and Crown forces from 1823 until 1852 were, without a doubt, of crucial importance, but none served fundamentally to threaten the Company's position in India. Despite this, commanding officers remained convinced that venereal disease could fatally undermine state stability. The apparent confidence of the colonial state from the 1830s to the 1850s masked the continuing fears of military and medical officials. Commanding officers and surgeons clung to the lock hospital system, despite its continued failure. Hamstrung by the fear of provoking discord amongst the soldiery, officials saw the system as the only viable option for containing the venereal menace.

Interestingly, this attitude changed very little following the rebellion of sepoy troops in 1857. If anything, the increased numbers of European troops in India following the rebellion further strengthened the conviction of pro-regulation surgeons and administrators of the need to reinstate the system.

This military-led mindset ensured that the lock hospital system was never far removed from official discourse. This chapter focuses on the period from 1831 to 1868, which began with the official abolition of the regulatory system in Bengal and ended with its formal return to favour. However, it suggests that the guiding principles of the system, as well as critical elements of it, never disappeared. Numerous commanding officers and surgeons worked around the abolition, quietly maintaining surveillance networks and utilising the newly opened government charitable dispensaries as surrogate lock hospitals. What follows highlights the intimate links that existed between the timing of the establishment of dispensaries and the closure of lock hospitals. Government dispensaries functioned as a mechanism to manage the health of those Indians seen as potential threats to the European soldiery. More critically for venereal disease control, surgeons attempting to cure these ills in their cantonments frequently co-opted government dispensaries and 'native hospitals' to serve these ends. The 'venereal wards' established at a number of the hospitals were important forerunners to the official re-introduction of dedicated lock hospitals. Despite this, the informal system of the 1830s and 1840s was no more successful than its formal counterpart. During the period of 'abolition', the Government of India was unable (or perhaps unwilling) to step in to censure the local decisions made by commanding officers and surgeons determined to uphold the system. Instead, they simply chose to look the other way. With echoes of earlier recalcitrant behaviour from army officers and provincial governors, Company and Crown officials subverted state policies, slowly chipping away, until the lock hospital system was formally re-introduced.

The dogged adherence to a system that had repeatedly failed to produce the desired results suggests that, for medical and military officers during the nineteenth century, the lock hospital system represented the dominant archetype of how venereal disease was to be dealt with, from which they refused to be moved, even in the face of overwhelmingly negative results. Surgeons and commanding officers continued to insist that *if only* the correct precautions were implemented, lock hospitals would not fail. This belief held European practitioners in India firmly in its thrall and only began to weaken in the early twentieth century, when the development of the arsenic-compound Salvarsan provided an actual cure for syphilis.

This chapter suggests that it was not just a lack of alternative medical options that bound this model so firmly in place, but continued underlying insecurity about the stability of colonial rule in India. The power this fear had in shaping military and medical decisions was most clearly demonstrated following Crimean war and 1857 Indian Rebellion when, in a relatively short period of time, the regulatory system was ushered back into official favour. The health (and fighting ability) of the army was once again framed as central to the preservation of British rule. Those governing India following the 1857 Rebellion were keenly aware of the deficits of the army. Officials attempted to assiduously avoid any 'preventable' risk such as syphilis, and the fear that venereal disease levels threatened the army's ability to mobilise and fight had once again become firmly entrenched. This chapter suggests the ways in which the royal commissions, formed after the 1857 uprising to investigate the organisation and health of the army in India, validated the demands made by the growing legion of surgeons and commanding officers to re-establish the lock hospital system. While 'experimental' lock hospitals had already been authorised at a number of stations, the publication of the *Report of the Commissioners Appointed to Enquire into the Sanitary State of the Army in India* in 1863 decisively opened the door for the expansion of the system on an all-India basis.

In her exploration of the Contagious Diseases Acts, Philippa Levine convincingly argued that the 1890s witnessed what she deemed a 'constitutional crisis.'[1] Tensions between the metropolitan and colonial governments over the Indian Contagious Diseases and Cantonment Acts threatened to erupt into a much more serious and destabilising threat for the government in London. Following the repeal of the Contagious Diseases Acts in Britain, an 1888 House of Commons resolution ordered the system's repeal in India. The resolution condemned both compulsory examination and the licensing of prostitutes.[2] Supporters of the system in India resisted all pressure to repeal the Act. In a move of seemingly outright defiance, in 1889 a new Cantonment Act (Act XIII of 1889) came into force that allowed for the forced expulsion from the cantonment of any 'diseased' individual refusing treatment or leaving hospital without permission.[3] The issue forced divisions not just between the Secretary of State for India and the Council of India (after the Council ignored the House of Commons directive on Contagious Diseases), but within the Viceroy's Council, as members threatened to resign over the issue.[4] This prompted the 'crisis', as resistance from

Indian officials, combined with the sensitive nature of the subject matter, nearly provoked huge political upset in London. As this chapter will show, this later friction resulted from the revival of an earlier technique utilised by surgeons and commanding officers during the period of 'abolition' from 1831 through the 1860s.

The involvement of anti-regulationist protesters and the growing nationalist movement in India further compounded the situation.[5] Indian officials – government members as well as medical and military officers – had come to closely associate regulated prostitution with the preservation of the health of the European troops, which in turn was connected to broader imperial security. This chapter illustrates the process by which the 'official' mind in British India came to associate the system of regulation so closely with the continuance of colonial rule. It argues that this must first be understood in order to explain how the situation reached its later crisis point. As Chapter 3 has shown, the formulation of a coherent discourse on venereal disease control in India gathered pace in the 1830s. This was fuelled by the various journals, societies and essays that focused the voices of surgeons and commanding officers and more effectively aligned their concerns. In the mind of its supporters, the lock hospital system took on a much greater meaning. It became not simply a means to control venereal disease and keep European soldiers healthy, but on a more subconscious level, the best (and in their view, only) means of approaching health and Indian bodies through a process that was itself emblematic of colonial control.

Responses to the closure of lock hospitals in the 1830s

The political and military battle over lock hospitals began almost as soon as Governor-General Lord Bentinck sanctioned their abolition in Bengal. As evidence to support his decision, Bentinck used the report of Inspector-General of Hospitals, William Burke, who suggested that not only had the lock hospitals failed to reduce venereal disease among the European troops, but in fact encouraged the reverse. Bentinck attempted to pressure Bombay and Madras to follow Bengal's lead in abolishing the system. This regional variance illustrates the continuing lack of a fully coherent hierarchy of control within imperial governance. This process was long and drawn out, as both Presidencies resisted Bentinck's pronouncement. John FitzGibbon, the Governor of Bombay agreed with its Medical Board in believing that lock hospitals should remain fully operational.[6] Seeking reinforcement, Bentinck complained directly to the Court of Directors about Bombay and Madras' stubborn refusal to comply with his request. At a time when cost concerns were all-pervasive, Bentinck pointed out that the suspension of lock hospitals would result in an immediate annual savings of Rs 30,000 for the Company.[7] Despite his pleas, however, both presidencies remained obdurate. Lock hospitals continued to operate in Bombay until 1833 and in Madras until 1835.[8] Moreover, at stations in Bombay where no civil hospitals existed to 'receive the diseased women', lock hospitals continued to function until their formal abolition in 1835.[9] Bombay grudgingly discontinued the 'compulsory system' only – deciding that civil hospitals would continue to admit those diseased women who 'voluntarily' applied for treatment.[10] In a similar manner, following the closure of lock hospitals in Madras, surgeons promptly sent women to the Native Poor Infirmary to continue their treatment.[11] The persistent refusal of both presidencies to bend to Bentinck's will demonstrates the widespread resistance to Burke's findings and orders from Bengal. Despite consistently lacklustre results and the fact that there was no direct military threat in either Bombay or Madras at the time,[12] both presidencies clung to the earlier conviction that venereal disease endangered the very stability of British rule in India. One Bombay Deputy Inspector of Hospitals warned menacingly that

> ... unless the Government takes some active steps to check it, syphilis will ere long prove a far more destructive disease than cholera itself. No disease more effectually renders a soldier unfit for the service and a burden to the state in a short time, or tends more to destroy his character for morality, sobriety, and good character [sic].[13]

These themes – morality, sobriety and good character – were woven into the fundamental understanding of how rule in India was maintained. Hence, if an external enemy did not attack a substantially weakened and unfit army, the more subtle, intimate threats presented by immorality, drunkenness and 'bad characters' would erode it from within.

Even within Bengal, surgeons and commanding officers defied Burke and Bentinck's decision. Burke's follow-up report, compiled two years after the closure of Bengal's lock hospitals, unwittingly revealed the steps taken by some commanding officers to 'supplement' venereal control measures in the wake of the system's abolition. Dinapur saw decreased admissions for venereal disease in the period following the abolition. On a visit to the station in 1831, Burke enquired as to the methods practiced by surgeons to produce this auspicious result. It soon became clear that Colonel Sole, the station's commanding officer, had, on learning of the impending lock hospital closure, arranged to hire the woman who worked as its 'matron' to '... watch over the condition of the prostitutes with whom the soldiers used to have intercourse.'[14] Her salary (paid out of regimental funds) was directly dependent on the level of disease among the European soldiers decreasing.

For its supporters, the lock hospital system represented not just the best means to treat venereal disease, but a broader symbol – a framework for how to deal with disease in India. Burke's identification of the failings of the coercive system fell, resoundingly, on deaf ears. While his supporters included Governor-General Bentinck and the cost-conscious Board of Control, his opponents remained a noisily vocal majority and proved the more resilient. The latter of these two sides would be the eventual victor. Not only was the system officially re-introduced in a relatively short period of time, but, by the end of the nineteenth century, the notion that disease control in India was necessarily an active, invasive (and in many cases, coercive) process, had become firmly entrenched in the minds of many surgeons and officials.

In 1831, Dr MacLeod, Bombay's Deputy Inspector of Hospitals, assured the Acting Adjutant General, Captain Keith, that the information he possessed indicated that venereal disease amongst the troops had been more prevalent before the existence of lock hospitals.[15] MacLeod attempted to make clear his belief that the reasons behind any failings in the system lay with the peons. Those supposedly 'inefficient' lock hospitals, he argued, were the ones *without* peons attached to detect and collect the women. As evidence, he noted that in Bombay from 1826 to 1829, when peons were not employed, the number of venereal admissions was 1234, but in the three subsequent years when peons were employed, admissions reduced to 647.[16] MacLeod's findings contradicted those of Burke, who asserted that there had been a decrease in venereal admissions among the men in Bengal since the lock hospital closure (see Table 5.1).

The reduction in cases, from one in every three men, to roughly one in every 4 1/3, Burke believed, was due to the fact that, after the closure of the lock hospitals, the women under their remit no longer fled the cantonment. The peons and inspections, as well as the prospect of being kept in the hospital, Burke observed, were hateful to the women, who responded by leaving the catchment area *en masse*. This exodus from the cantonment had a knock-on effect. Soldiers resorted to seeking out the women, traipsing some distance from the cantonment and exposing themselves to all weathers, day and night, to find them. As a result, the men were more likely to be afflicted with fevers, dysentery and cholera.[17]

Table 5.1 General abstract of venereal diseases in His Majesty's regiments in Bengal – from 21 December 1825 to 31 December 1831

	Years					
Venereal diseases	**1826**	**1827**	**1828**	**1829**	**1830**	**1831**
	Strength of regiments in the command					
	6,874	**8,760**	**8,812**	**8,315**	**8,914**	**8,898**
Syphilis	509	663	504	757	272	337
Ulcers penis non syphiliticum	126	450	734	603	385	545
Bubo simples	142	498	504	407	393	365
Gonorrhoea	368	719	807	629	637	622
Hernia humoralis	130	215	199	153	190	186
Total venereal diseases	1,275	2,545	2,838	2,549	1,891	2,055
Proportion of venereal diseases to strength of command of His Majesty's regiments	1 in 5¼	1 in 3½	1 in 3 1/10	1 in 3¼	1 in 4 2/3	1 in 4 1/3

Source: Letter to Colonel Casement from W Burke, 21 April 1832. APAC, F/4/1338/53031.

Burke argued that with the abolition of the lock hospitals, a 'better class' of prostitute once again returned to the bazaars and cantonments, who was no longer afraid of European military hospitals and therefore willing to voluntarily apply for treatment when diseased.[18] Burke's explanation, however, remains unconvincing, as it was not just the inspections and the peons that were detested, but the treatment itself and the enforced stay (during which time, the women could not earn money) at a hospital that was often some distance from family and friends.

What then, did Burke propose to replace the system? He suggested that dispensaries and military hospitals could function as a substitute to the lock hospital system. Burke recommended that the surgeons of each regimental hospital be given the power to administer medicines to all diseased women who applied.[19] Burke asserted that once 'public'[20] notice was given to this effect, positive results were sure to follow. However, these proposals must have appeared at once vague and naïve to those like MacLeod, who believed that the system's success was only possible with compulsion.

Burke died a short time after the release of his follow-up findings. However, the battle over lock hospitals continued. After his death, a number of reports were produced for the Court of Directors on the subject of venereal disease. The recommendations of these reports never failed to conclude that the re-establishment of the lock hospital system was urgently required. In 1838, following a recent dramatic increase in venereal admissions at some stations in Madras, John Murray, Deputy Inspector General of Hospitals, at the request of the Commander in Chief, addressed a circular to all medical officers responsible for Crown regiments in the presidency.[21] This circular functioned as yet another platform for surgeons to argue for the return of the regulatory system. Like the 1831 Bengal venereal disease circular, Murray prompted his surgeons for their personal experiences in various treatment methods. He asked each of them whether they had

> ... reason to believe that this [venereal] class of disease has increased, or decreased, or on average continued in the same ... since the abolition of Lock Hospitals in 1835, as compared with its degree of prevalence for some years prior to that period ... [further] whether or not the nature of these diseases has become more virulent since 1835, than previously.
>
> I request at the same time that you will endeavour to find out whether syphilis and gonorrhoea have become more prevalent of late years ... [and] what steps, if any are taken to cure the Native Women when discovered to be infected; and further, that you will give me your opinion as to the best mode of checking the frequency of these diseases among the soldiery.[22]

Murray's questions about prevention can be read as an invitation for surgeons to proclaim their support for the lock hospitals. In view of the replies he received, Murray could not have been disappointed. These reveal a fervent espousal and enduring attachment to the lock hospital system among regimental surgeons in Madras. Each of the replies that Murray included in his report to the Medical Board suggested the re-establishment of a 'reformed' lock hospital system as a matter of some urgency. In his covering letter to Major General Fearon of the Military Board, Murray boldly pronounced that the 'experiment' of discontinuing the lock hospitals had singularly failed and the time had now come for this great wrong to be rectified.[23] We should not dismiss this phrasing as simply an impassioned polemic. Murray and the surgeons who responded to his circular saw the lock hospital system as standard and its absence an aberration. They could not conceive of a way in which medical men in India could ensure the health of the men under their care *without* the 'protection' promised by the lock hospital system.

However, like Burke's figures, those that Murray included in his report are difficult to interpret and seem to prove neither the success nor failure of the lock hospital system (see Table 5.2).

Lock hospitals had remained in force in Madras until 1835 while venereal cases had already begun their rise well before that time. What both sets of figures suggest is that levels of venereal admissions fluctuated constantly, seemingly unaffected by the opening or closure of lock hospitals and contingent instead on a range of other, unexplored considerations. In seeking the reasons behind the jump in admissions observed in 1833, it is perhaps more important to look at the other factors which might have influenced the health of those in the cantonments. For instance, disease levels tended to fluctuate when armies were engaged in campaigns (as was the case for those Madras regiments engaged in suppressing rebellions in Vizagapatam between 1832 and 1833).[24] Instead

Table 5.2 Comparative state of venereal diseases treated in Her Majesty's regiments in the presidencies of Bengal and Madras from 1 January 1830 to 30 June 1838

Year	Bengal presidency				Madras presidency			
	Annual average of strength	Total of venereal diseases	Proportions of venereal to strength	Percentage of ditto	Annual average of strength	Total of venereal diseases	Proportions of venereal to strength	Percentage of ditto
1830	8914	1891	1 in 4⅔	21.21	4317	293	1 in 14¾	6.78
1831	8898	2055	1 in 4⅓	23.09	5068½	506	1 in 10	10
1832	7872	1584	1 in 5	20.12	5470⅓	1172	1 in 4⅔	21.46
1833	7424	1182	1 in 6⅓	15.92	5453⅔	2436	1 in 2½	44.67
1834	7707	1341	1 in 5¾	17.40	5921	2481	1 in 2½	41.96
1835	8181	1440	1 in 5¾	17.60	6383	2282	1 in 2¾	35.75
1836	7307	1614	1 in 4½	22.09	6707¾	2278	1 in 3	33.95
1837	8058	1521	1 in 5¼	18.87	6487¾	2255	1 in 2¾	34.76
1838 for 6 months	6876	841	1 in 8½	12.23	5842⅙	1097	1 in 5⅓	18.77

Source: Letter to the Secretary to the Government of India, Military Department from Mr Steel, Secretary to Government, Fort St George, 6 November 1838. NAI, Home (Military). Consultations, Nos. 70–72, 4 February 1839.

of seeking such alternative explanations, however, surgeons saw the system as the salient feature in determining venereal admissions.

The response of Surgeon Servis of the 4th King's Own to Murray revealed another important facet to the lock hospital system's endurance. He noted that although he had *no direct experience* of the official lock hospital system (having only arrived in the country eight months prior), he had it on *good authority* that disease had increased considerably since the system's abolition.[25] Servis had come to this conclusion through the enquiries he made with nearby 'competent medical observers'. Servis was not unique in how he came to form his opinion about supposed 'best practice' in treating venereal disease. The proponents of the lock hospital system had a much more efficient and dedicated propaganda machine in place than their opponents. Throughout the 1830s and 1840s, in an almost evangelical fashion, a relatively small group of surgeons systematically 'educated' younger surgeons as to the 'essential' nature of the lock hospital for maintaining European health in India. This 'education' was accomplished not just through word-of-mouth and periodic reports, but as Chapter 3 argued, took the form of published articles and presentations to the medical and scientific societies.

Servis' letter also exposes the continuation of punitive methods of control exercised over women deemed to be prostitutes. He notes that 'Measures have been *constantly* taken by the Commanding Officers, by means of the constables, to capture these women [believed to be diseased] but without much success.'[26] Even without the physical presence of a lock hospital, aggressive measures to contain women in and around the cantonments remained firmly in place. Some surgeons, like the one in charge of the 55th Regiment in Madras, went even further; in 1838 he ordered the establishment of a temporary lock hospital.[27] This move was in keeping with that of the majority of surgeons who supported the appointment, at the very least, of a 'medical police' to ensure the periodic examination of all suspect persons.[28]

Surgeon Mouat of HM 13th Dragoons at Bangalore was one of the few surgeons who sought answers to the problem of continually high venereal admissions *outside* the commonly accepted framework (which repeated three excuses for the failure of the system to accomplish its goals: the improper functioning of the lock hospital; the duplicity of the women or peons; or the lack of sufficient methods of detection). Mouat instead asserted that one could conclusively link variations in levels of disease to periods of famine and bounty.[29] He came to this conclusion by analysing admissions to hospital during the supposed years of 'plenty' and those of 'famine' Mouat's reasoning was that poorer women were driven by starvation to sell themselves, or, alternatively, parents their daughters, in exchange for money to buy food. By his reckoning, the years 1820–1832, 1826–1832 and 1835–1837 were famine-free, while 1824–1825 and 1833–1834 saw widespread famine. During the years of 'plenty', there were 78 admissions among the Dragoons for gonorrhoea and syphilis, while during the years of 'dearth', admissions rose to 216.[30]

While it is interesting that Mouat sought alternative explanations for disease fluctuations, he nevertheless continued to link spikes in disease to the behaviour of poor native women. His argument suggests that indigent women sold into prostitution (who, according to the logic of his argument, were not previously prostitutes), were responsible for increases in disease levels. Yet he does not take the step of suggesting *who* infected these newly initiated prostitutes. Mouat's argument once again frames Indian women (here, poor and famine-ridden) as the vectors of disease. While he blamed those women who felt compelled to sell their bodies or face starvation, no corresponding critique was made of Company policies that exacerbated famine conditions.[31] Nor did he lay out any proposals to protect against future famines. While this lack of criticism towards government is hardly surprising, it highlights the pervading single-minded determination to situate blame for venereal disease with Indian women; wilfully ignoring any suggestion of the role played by either soldiers or the policies of the Company.

Mouat then goes on to attack Burke's report, noting that it '... merely gave results without tracing them to their causes'.[32] Burke's failure to look at anything but the starting and end points had led, Mouat argued, to the misguided and dangerous calculation which resulted in the closure of lock hospitals. However, in suggesting the reasons behind the problems which existed 'in between', Mouat once again fell back on a familiar trope – that of the deficiencies of the 'native character.' Therein, he argued, '... lies the whole mystery and on it the onus of the failure of the Lock Hospital system.'[33] The generalisations that followed included many of the well-worn arguments expressed by earlier European surgeons and officers. These included the supposed Indian propensity towards greed and 'rascality' believed to be endemic, as well as the 'cupidity' of the peons hired to enforce the system.[34] Mouat asserted that one could locate the reasons behind the failures not only of earlier lock hospitals, but of *all* European institutions in India in native subterfuge and avarice. He asked

> Do we not find every institution and almost every regulation by native craft and artifice in this country a source of profit? Thus once

> established, it is easily seen how the increase of disease must tend to the increase of pecuniary advantage. It thus spreads and it is manifestly their interest that it should increase.[35]

This became one of the most frequently cited reasons for earlier failures. Surgeons laid the blame for compulsion and corruption squarely at the feet of Indian intermediaries – the peons, *chupprassies* and matrons who, ironically, were at the same time hailed as integral to the functioning of the system. The Officer Commanding the Mysore division, in a letter to the Quarter Master General, insisted that, '... these functionaries made use of their power for extortion and all sorts of villainies,' proffering countless bribes not only from 'common prostitutes' but from 'respectable families'.[36] These accusations were something of a double-edged sword, attached as they were to the assertion that such coercive functionaries were necessary for the system's proper operation. They represent what Levine has called the uncomfortable 'ambivalence between a Macaulayite' ideal of well-trained 'brown Englishmen' and the fears of 'flawed assimilation.'[37]

Citing such villainies, Mouat did not seek a re-establishment of the system on identical lines as it had previously operated, but advocated what he deemed a 'modified' version. However, once again, Mouat's suggestions represented a lock hospital in all but name. He proposed the establishment of a 'native hospital' to treat infected women. A matron would be placed on the staff of this hospital who would monitor all local 'females of pleasure' and 'induce' them to go to hospital when diseased or apply for their expulsion from the bazaar if they refused.[38] Little, save the name, was altered by these plans. Despite the seemingly transparent nature of this argument, surgeons frequently repeated similar semantic manipulations over the course of the next 20 years which saw lock hospitals designated instead as 'bazaar hospitals' or 'hospitals for diseased women'.[39]

The Medical Board, for its part, while inclined to agree with many of Mouat's assertions, rejected his proposals to re-engineer 'native hospitals' to suit venereal needs. Instead, it sought a middle ground and proposed the establishment of infirmaries and general hospitals for *anyone* labouring under venereal disease.[40] While outright compulsion was prohibited, it was understood that these infirmaries and dispensaries would fill the space left by the closure of lock hospitals. The Governor-in-Council agreed with the Board's plan for dispensaries (tellingly, only at stations where European troops were stationed), but refused to approve the proposals for matrons to force women into treatment.[41]

However, with all the other tools remaining in place, it seems likely that compulsion and forced treatment continued at many stations. In the case of venereal disease control, the will of the civil authority was (at best) ignored or (at worst) actively subverted by the Company's own *British* employees.

Compulsion did not simply take the form of forced examination and treatment. Here again we find further evidence on the degree to which the ideas that moulded the lock hospital system had become ingrained in the minds of commanding officers and surgeons. As discussed in Chapter 4, commanding officers frequently expelled 'unwanted' or troublesome women from the cantonments to protect the health of the soldiers. While as late as 1852, the Governor-General was unwilling to authorise such a power, noting that he did not feel it 'practicable to legislate with advantage' these measures,[42] expulsions were authorised in areas where local magistrates or commissioners were sympathetic to military demands. One such mass eviction took place in Hazareebagh[43] in response to a plea made by the Adjutant General to clear the civil bazaars of women believed to be diseased.[44] The Deputy Commissioner responded by ordering the removal of all 'kusbees, touifes [sic] or dancing women' from the bazaars.[45] Here, it is interesting to note that not only was the Commissioner willing to bend to military demands, but that he also embraced military understandings of the category of 'unwanted' 'prostitutes'. In a similar manner, while Adjutant General Lieutenant Colonel Alexander of the Madras army professed his disapproval of the lock hospital system, he nevertheless suggested that *all* cantonment residents be subject to regulations administered by a specific Military Police.[46] However, Alexander's 'all' implied only Indian civilians, rather than European personnel. Such a force would be empowered to expel all 'bad characters' from the cantonment. Captain TR Morse, commanding at Poona called for a similar 'sanitising' power to force all 'kusbins [sic] and wandering people who live in huts and hang about the city and cantonments' to be banned from coming within 10 miles of the cantonment.[47] Morse described these individuals as 'the pest of the place – professional thieves and nuisance of the greatest magnitude.'[48] The Board not only approved of Morse's plan, but ordered that a letter be sent to the Judicial department, instructing the local magistrate to cooperate fully with the military authorities to ensure the smooth running of the plan.[49] The Judicial department not only complied, but expanded on this order in 1851, recommending the adoption of strict measures to remove 'objectionable characters' from the area around military cantonments where European troops were stationed.[50]

The fact that such orders included *tawa'ifs*, 'kept' women and 'wandering peoples' again indicates that authorities not only regarded these women as 'prostitutes' but had abandoned any earlier reluctance to offend 'native prejudices' as they launched these offensives. Moreover, it highlights the broader targeting of individuals deemed to be a threat to the state, such as those who refused to be 'settled'. The author of an 1863 letter, describing the recent history of lock hospitals, casually mentioned that from 1850, military authorities stressed the need to prohibit 'kulatnees' (described by the author as a 'race of thieves whose women are all prostitutes') and vagrants from settling near military lines as both were seen to be responsible for the spread of venereal disease among the soldiery.[51]

Few officers remained alert to the charge that such institutions might offend 'native sensibilities' and a handful addressed this in their reports. Yet, the terms in which they phrased their justifications were now rather different from those made in the 1810s and 1820s. The understanding now was that 'upright' Indian society had shunned such 'undesirable' woman as 'common prostitutes'. The Medical Officer at Secunderabad, in his attempts to reassure government that native society would not be greatly disturbed by the re-emergence of the lock hospitals, stated that he was confident that 'the most respectable part' of the native community was unlikely to oppose further measures to implement a full registration and inspection system as such a system would bring great public advantage.[52] In at least one respect, it is possible that the medical officer wasn't entirely mistaken. While nationalist groups were actively involved in the protests against the Contagious Diseases Act, the fundamental categorisation of women as 'fallen' or 'prostitutes' was, by this time, rarely questioned. The 'inner' domain that Partha Chatterjee has described was used by nationalists to identify their difference (and moral and spiritual superiority) to the European colonisers[53] did not include women deemed sexually 'deviant'. The battle nationalists waged against the Indian Contagious Diseases Act was formed on the basis of its reinforcement of a system of 'legalised vice' which was only interested in preserving the health of the European soldier. The first recommendation passed by the Indian National Congress at its third session in 1887 demanded the complete repeal of all legislation that related to the regulation of prostitution.[54] However, neither nationalists nor external European observers questioned the categorisation of the women as 'fallen'.

The perceived threat to colonial rule posed by venereal disease was never far from official discussions. In April 1841, Governor-General

Auckland requested the Medical Board provide him with a detailed report outlining the measures taken to prevent such diseases. Auckland was aware of the suggestions of surgeons for the control of venereal disease. In particular, he wanted to know if the measures proposed by Dr Playfair (the Superintending Surgeon whose circular and report on venereal disease is discussed in Chapter 2) were now practical.[55] In reply, the Medical Board sent Auckland various representations in support of the lock hospital system (prominent among which was a strongly worded letter from Murray). The specific fears held by medical and military officers regarding venereal disease now tugged at the insecurities of governors, advisors and European observers. The plans for 'public health' which emerged in the aftermath of lock hospital abolition clearly highlight this.

The dispensary and charity hospital

How did the new dispensaries impact the women living in and around military areas and what were the broader implications for 'public' health? There was a close correlation between the demise of the lock hospital and the development of charity hospitals and dispensaries; the establishment of the latter coincided almost exactly with the closure of the former.[56] Many surgeons, seeking alternatives to the abolished regulatory system, quietly turned to dispensaries and 'native hospitals' to deal with the continuing problem of venereal disease amongst the soldiery. These new institutions represented a continuation of military health by stealth – ostensibly operating for the 'public health', many nevertheless remained at the mercy of army demands.

The first native dispensary to open in Bombay did so in 1834.[57] The dispensary was located between the Fort and the native town, funded in large part by the government. Three kinds of dispensaries emerged at this time: those where the cost was borne by the community; those where the cost was shared between government and the community; and those where the cost was borne entirely by government.[58] It was this last category that would be most frequently (and easily) utilised by military and civil surgeons following the abolition of lock hospitals. Charity dispensaries or hospitals funded by wealthy Indian patrons, like Bombay's Jamshedji Jeejeebhoy hospital (built in the 1840s), could not be so easily railroaded into serving the military's needs.

While the bare-faced methods of coercion utilised by the former system were officially denounced, many commanding officers continued to rely on an informal 'matron' system, in combination with the new dispensaries, in their attempts to control disease. Although in its 1838 report, the Madras Medical Board was careful to point out that it did not support compulsion, in the suggestions for the establishment of 'benevolent' dispensaries, the Board notes that the police should expel those women who refused to agree to medical treatment from the regimental bazaars.[59] The Board also continued to support the matron system, attaching it now to dispensaries, rather than lock hospitals, but still making the woman's pay contingent 'on the value of her services' in reducing levels of venereal disease among the European troops.[60] The Board made this link despite the fact that it was this very same 'incentivisation' that had led to earlier claims of deception, compulsion and bribery.

In Secunderabad from 1844 until 1848, the Superintendent of Police revived another controversial practice related to the lock hospital system – that of taxing women who plied their trade in the cantonment. He ordered a levy imposed on all 'Dancing Girls, Cottage Women and Prostitutes' to pay for the salaries of two matrons and one peon to monitor and patrol the women.[61] Each woman was required to report any suspected venereal disease to the matron, who would then provide her with medicine from the hospital. Any woman found to be 'carrying on commerce' in a diseased state and discovered was 'severely punished'.[62] At this station, then, not only were women forced into the system, but were made to pay for their own suasion. Moreover, the range of women included under by this net continued to expand.

In addition to the appropriation of dispensaries for military needs, an 1841 letter to the Calcutta Medical Board revealed still more explicit transgressions of the 1831 order. The hospital returns for Benares in 1840 ranked primary syphilis as second only to 'cutaneous disease' in the number of patients admitted.[63] In response to this, the following year, the Governors of the Benares Native Hospital ordered it converted to accommodate a lock and police hospital on its grounds.[64] The Governors of the Hospital, clearly keen to protect Benares from the venereal menace, made a 'special provision' for suffering prostitutes. Those women deemed to be incurable, or thought to require lengthy treatment, were sent to the Blind Asylum for sustained (and presumably restrained) treatment.[65]

While treatment for diseased women took place largely in the city or charity hospitals, in a few instances we find examples of hospitals and dispensaries established solely for bazaar residents, such as the Sudder Bazar Hospital at Akyab.[66] However, returns from this period often list a separate category of patients not seen in earlier returns. These were the 'house' patients, treated by surgeons in their own homes.[67] It seems likely that some of these patients were the women formerly under the remit of the lock hospital. In 1845, Mr Steven, forwarding his half-yearly report on the Muttra and Agra dispensaries, when discussing the house- and out-patients under his care specifically highlighted increased levels of syphilis in Agra.[68]

In his 1848 suggestions for cantonment 'cleansing', Captain Morse proposed that those bazaar prostitutes who 'voluntarily' submitted to a regulatory system be allowed to remain in the cantonment. To monitor these women, Morse argued that bazaar funds be used to pay for the employment of one matron, one 'muccadum' (or head man), seven peons and one apothecary to establish a system whereby the women would be periodically examined and treated for disease *without* recourse to a lock hospital. Morse proclaimed that the women could have no cause for protest as they would be examined and treated in their own homes.[69] In 1849, Brigadier Boileau reported that venereal disease had decreased among the European men under Morse's arrangements, but, due to the great expense of treating the women in their own homes, a 'new' plan had been devised, which Boileau referred to as a 'House Dispensary' whereby the women were paid subsistence during their stay at the dispensary.[70] However, in 1850, Morse reported that after two years, his plan had failed. He stated that he could not see a way for the plan to succeed *unless* government authorised the extension of police supervision on a massive scale. Notwithstanding Morse's admission of failure, the system was continued as a 'trial' for two more years, after which the government again declared the system to be 'inefficacious'.[71] Nevertheless, despite this admission, government continued to sanction the expenses incurred for 'relieving the diseased'.

In Trichinopoly and Bellary, disease levels among the European troops were so serious that female venereal dispensaries and wards were authorised in 1844.[72] Initial reports from the ward in Trichinopoly stressed that every attention was taken to 'conciliate' the women by taking great care with their wants and comforts.[73] Returns listed both in- and out-patients, suggesting again that surgeons treated some women in their homes.[74] However, it would appear that unofficially, methods of compulsion remained in place. The surgeon compiling the report admitted that '... some of the inmates of the hospital, remained there more under the dread of compulsion than from their own free will.'[75] He went on to acknowledge that the women (except in the most extreme cases) were unlikely to come on their own volition, but he defended the pressure applied to them by saying that the 'only' compulsion used was the threat that those who refused to come voluntarily for treatment would be '... Tom Tomed and expelled from the lines.'[76]

Like the lock hospitals before them, few positive results emerged from Trichinopoly and Bellary's venereal wards; after four years of operation, there had been very little reduction in disease levels among the men. Instead, the reverse seemed true at Trichinopoly, where higher rates of both gonorrhoea and syphilis were recorded during the four years of the ward's operation than in the five years prior (See Table 5.3).

Dispensaries also functioned as places of experimentation. Not only did they represent some of the first significant forays into what could be called 'public health' in India, but they were the spaces where young doctors experimented with new methods on their (in the case of many bazaar dispensaries and jail hospitals) captive patients.[77] Reports from the Government's Charitable Dispensaries in Bengal in 1849 reveal a number of patients treated by the (then-popular) method of mesmerism.[78] By this method, patients would be 'mesmerised' and while in this hypnotised state, procedures and surgeries performed. At Hoogly,[79] where James Esdaile established his mesmeric hospital, the dispensary records the case of a 20 year old woman named Shama (identified only as a 'prostitute') in May 1849. Shama had a sarcomous tumour roughly the size of a grapefruit on the left side of her vagina. The report gives no further details on her, only on the experiments conducted while she was under hypnosis. For example, the report notes that while mesmerised, she was '...indifferent to pricking and pinching' and that the following day, the '...tumour was exposed and severely squeezed.'[80] Shama was mesmerised every day for four days and on the fourth surgeons excised and removed the tumour.[81]

It is not surprising that at these dispensaries, aligned as they were (in the minds of many surgeons) with the primary objective of preserving European health, some degree of bullying was employed to 'persuade' Indian patients to attend. The Court of Directors acknowledged this in their discussion of Madras dispensaries, noting that '... in certain cases [coercion has] been used to induce persons to attend.'[82] While the Court warned surgeons and commanding officers that this practice was best avoided as it engendered a 'prejudice' against these institutions, the wording of the warning suggests that it was not the general population whom the Court had in mind, but the would-be former lock hospital inmates. Moreover, the Court went on to temper its reproach with the admission that in cases where individuals had the potential to communicate disease to others as a result of their non-attendance at the dispensaries, some deviation from this rule was permissible.[83] With such potential loopholes, those surgeons and commanding officers who were determined to maintain some form of regulatory system

Table 5.3 Number and percentage of admissions from Syphilis Primitiva, Gonorrhoea and Hernia Humoralis among the European soldiery at Trichinopoly and Bellary from 1840 to 1848, inclusive

					Trichinopoly					
	1840	**1841**	**1842**	**1843**	**1844**	**1845**	**1846**	**1847**	**1848**	**Total**
Strength	620	415	462	877	1052	948	914	929	1093	7310
Syphilis primitiva	35	57	46	62	28	59	56	64	66	473
Gonorrhoea	35	107	30	58	62	106	100	116	87	701
Hernia humoralis	7	10	8	39	13	25	38	41	24	205
Total	77	174	84	159	103	190	194	221	177	1379
Percentage of treated to strength	12.42	41.93	18.18	18.13	9.79	20.04	21.23	23.79	16.19	18.86
					Bellary					
	1840	**1841**	**1842**	**1843**	**1844**	**1845**	**1846**	**1847**	**1848**	**Total**
Strength	892	642	602	896	1015	1123	481	989	1125	7765
Syphilis primitiva	206	143	112	92	119	158	46	182	141	1199
Gonorrhoea	181	89	93	122	134	218	136	230	211	1414
Hernia humoralis	26	21	13	33	29	29	10	26	24	211
Total	413	253	218	247	282	405	192	438	376	2824
Percentage of treated to strength	46.40	39.41	36.21	27.57	27.78	36.06	39.92	44.29	33.42	36.37

Source: Letter to JF Thomas from Assistant Surgeon A Lorimer, 5 July 1849. APAC, F/4/2341.

had little trouble doing so without prompting any serious reprimand from government. By the late 1840s, something resembling the original lock hospital system had begun to re-emerge as dispensaries and charity hospitals were co-opted and 'experimental' venereal wards established.

Working around the abolition

Momentum for the official return of the lock hospitals (by whatever name they were given) gathered pace in the 1850s. Perhaps realising this, commanding officers and surgeons tested the water to see what they could get away with. Superintending Surgeon Young, commenting in 1849 on the recent rapid increase of venereal diseases among the men of HM 94th Foot in Malabar and Canara, ordered one of his assistant surgeons to coordinate with the garrison surgeon to set up separate 'venereal wards' for the women deemed to be the highest threat to the soldiers. Young estimated that there were 37 venereally-infected women in the Cannanore bazaars.[84] That he had this figure so readily to hand suggests not only that surveillance networks had remained in operation in Cannanore, but that internal examinations were still taking place. In his endeavours, Young had the support of the Commander of the 94th Regiment, Lieutenant Colonel HR Milne. In November 1848, Milne forwarded a list of the 37 women to the Major of the Brigade at Cannanore (see Table 5.4).

This list was, in effect, an unofficial census of Cannanore's cantonment prostitutes. The women were asked to state their name, place of origin, current residence and whether they were willing to return home.[85] Young and Milne then combined these answers with the results of the women's internal examinations.

It is difficult to determine the accuracy of Milne's list (not least for the contentiousness of the 'prostitute' categorisation), but what is clear is that, with 26 of the 37 women reported as 'sick', Milne believed that urgent steps were required to counteract the rising venereal threat. This list begs a number of further questions. The 'native place' given of many of the women reveals that a number of them had travelled a substantial distance to Cannanore. While there is no additional evidence provided to suggest the reasons why they made their journeys, it is possible that at least some of the women had initially arrived as the companions of soldiers or officers (or, more generally as camp followers), travelling from one military site to the next. Reports on 'dissolute' European women suggest that many women often sought the 'protection' of another man after their own husband had either died or abandoned them. This

Table 5.4 Names of common prostitutes residing in the cantonment of Cannanore, 17 November 1848

	Names	If sick	Where residing	Native place	If willing to return home
1	Jane Margrate	Sick	Government hospital	Bombay	No
2	Poolam Murriah	Sick	Tallook road	Quilon	Yes
3	Patmoor Kirby	Sick	Tallap road	Quilon	Yes
4	Fantmeer	Not able to walk	Pension line	Quilon	Yes
5	Lutchmeer	Sick	Pensioners line	Quilon	Yes
6	Yantee	Sick	Pensioners line	Trichinopoly	Yes
7	Pavolee	Sick	Camp bazar	Quilon	Yes
8	Allimell	Sick	Camp bazar	Bangalore	Yes
9	Muniah	Sick	Butcherers place	Cutcheen	Yes
10	Lutchmee	Sick	Butcherers place	Manglor	Yes
11	Sessillia Decist	Not sick	Camp bazar	Cannanore	Yes
12	Murriah Rosier	Not sick	Camp bazar	Calicut	Yes
13	Coolie Madue	Sick	Chucklers lane	Tellicherry	Yes
14	Emmasa	Sick	Camp bazar	Cannanore	Yes
15	Murrier	Sick	Moholy hind gates	Madras	Yes
16	Emmasser	Sick	Regimental bazar	Cannanore	Yes
17	Murriah	Not sick	Regimental bazar	Cannanore	Yes
18	Moolmah	Sick	Regimental bazar	Pallincoolee	Yes
19	Mummiah	Not sick	Regimental bazar	Cannanore	Yes
20	Collin	Not sick	Regimental bazar	Tellicherry	Yes
21	Voulie	Sick	Regimental bazar	Cannanore	Yes
22	Tommy Carly	Sick	Regimental bazar	Tellicherry	Yes
23	Ammamah	Not sick	Regimental bazar	Cannanore	Yes
24	Tooncamah	Sick	Regimental bazar	Madras	Yes
25	Rossier	Sick	Regimental bazar	Cannanore	Yes
26	Maghan	Sick	Regimental bazar	Madras	Yes
27	Pooniar	Sick	Regimental bazar	Madras	Yes
28	Mummiah	Not able to walk	Regimental bazar	Madras	Yes
29	Tenengor	Sick	Regimental bazar	Calicut	Yes
30	Tommae	Not sick	Camp bazar	Cannanore	Yes
31	Pascole	Not sick	Regimental bazar	Cannanore	Yes
32	Espallo	Sick	Regimental bazar	Cannanore	Yes
33	Fimerly	Not sick	Regimental bazar	Madras	Yes
34	Eliza	Not sick	Regimental bazar	Quilon	No
35	Collanah	Sick	Regimental bazar	Madras	Yes
36	Petteche	Sick	Regimental bazar	Madras	No
37	Terreaser	Not sick	Regimental bazar	Cannanore	No

Source: Letter to the Major of the Brigade, Cannanore, from Lieutenant Colonel HR Milne, Commanding HM 94th Regiment, 19 November 1848. Madras Military. Board of Commissioners, Collections, 1849–1850. APAC, F/4/2341.

might have been the case for Jane Margrate, one of only four women on the list who stated their unwillingness to return to her 'home' and the only woman specifically identified as a 'European' (if we apply Durba Ghosh and Indrani Chatterjee's assertion that the presence of a surname in the colonial records is an indicator of a woman's European status).[86] Moreover, at least two of the women have names which could suggest Eurasian origins: Eliza and Terreaser (Theresa), who were similarly unwilling to return 'home', but who, unlike Jane, were not diseased.

In forwarding this list, Superintending Surgeon Young begged for Commissariat funds to hire a *chowdrannee* and two washerwomen in conjunction with the venereal wards.[87] Without waiting for the reply of government, he set up an 'emergency' lock hospital through 'private subscription' – in this case, a collection from the officers of the regiment.[88] Like many surgeons before him, Young pressed the Medical Board to ensure that the civil authorities cooperated fully with both medical and military authorities.[89] He requested that the Collector of the District and the Superintendent of Police support him in giving the *chowdrannee* their full assistance in capturing diseased women. This was clearly a concern as Young believed that the limits of the cantonment were insufficiently defined, and as such, might permit diseased women to evade military jurisdiction by applying to the civil authorities for protection. Instead of reprimanding Young for this action, however, the Madras Government approved his establishment, calling only for the submission of a follow-up report on its progress.[90]

By 1848, there was a groundswell of support for the reinstatement of the lock hospital system. In an extensive report to the Deputy Adjutant General, Lieutenant Colonel Havelock, Bombay's Deputy Inspector of Hospitals Surgeon Kinnis, laid out his observations on venereal disease in that presidency. Kinnis argued that Burke was very much mistaken in his assertion that since the abolition of the lock hospitals, the proportion of venereal disease had decreased.[91] Instead, Kinnis attempted to demonstrate that from 1826 to 1836, during the operation of lock hospitals in Bombay, the proportion of venereal admissions to strength was 23.28 per cent. In the years since abolition (namely 1837 to 1847), this number had risen to 26.29 per cent.[92] Perhaps aware that critics might dismiss a 3 per cent rise as insignificant, he went on to attack Burke's accounting methods. In particular, he criticized the way in which Burke compared venereal admissions in 1826 and 1831. Burke, Kinnis argued, incorrectly measured the figure for 1831 (a full year) against that of 1826 (a six month period).[93]

Despite vociferous proclamations to the contrary, the system re-established after 1850 was almost identical to earlier incarnations. Hailing the return of the lock hospital as something of a moral victory, surgeons and commanding officers once again extolled the charitable and benevolent aims of the institutions and the benefits they would bring to poor, dissolute women – their would be inmates. Yet, the real reason behind their re-establishment remained clear and constant- to protect the health of the European soldiery. In 1855, due to the 'fearful' rise of venereal admissions among European soldiers in Bangalore, lock hospitals were authorised on a 'temporary' basis.[94] The establishment of lock hospitals at Bangalore and the sanctioning of broader rules governing them by the Madras government in 1860 soon prompted the establishment of six further lock hospitals at Secunderad, Bangalore, Cannanore, Trichinopoly, Wellington, Bellary and Vizagapatam.[95] By 1862, five more hospitals had been authorised (but not yet constructed) at Rangoon, Kamptee, Poonamallee, Saint Thomas' Mount and Jaulnah.[96]

The alarm over increased levels of venereal diseases in the 1840s had proved to be the thin end of the wedge in gradually encouraging the re-emergence of regulatory systems in Indian cantonments. However, it was the heightened fears surrounding army health that followed the Crimean war and 1857 uprising, combined with the findings of the multiple sanitary and military commissions, which finally heralded the official re-appearance of lock hospitals. The movement for re-instatement, begun by surgeons like John Murray in Madras and carried forward by numerous others across the subcontinent, had by the 1850s, secured a major victory. By consistently arguing that, with proper 'aids', lock hospitals could be successful, campaigners ensured that both medical boards and government now regarded Burke's findings with scepticism.

Wars and sanitary commissions

Following the catastrophic loss of life attributed to medical failures during the Crimean war and 1857 Indian rebellion, parliament launched campaigns and commissions to investigate every aspect of the army – from its organisation to the construction of barracks to the health of the corps. Among those who gave evidence to the 1858 Commission on the Organisation of the British Army in India were a number of European surgeons who warned of the dire state of the health of the troops. John McLennan, recently retired from the position of Physician-General of Bombay, warned that the rampant and widespread nature of venereal disease in India demanded swift and efficient police regulations.[97] Excepting incidences of cholera, McLennan claimed that 50 to 70 per cent of all admissions to European hospitals in India were due, directly or indirectly, to syphilis or drink.[98] He suggested barring all diseased 'public women' from living within five miles of cantonments. However, he took this argument further, stating that 'All wandering tribes, Gipsies, Kolalnees, Nuths, &c, whose women are of the above character, should (the women I mean) be punished, if they were found nearer than 5 to 10 miles' unless they agreed to register themselves and submit to inspection.[99] This move towards forced expulsion should again be viewed in light of the continued reluctance to restrain or castigate the men for their behaviour. Forcing the women 10 miles from the cantonment ensured that the men could not easily reach them in the course of their daily 'wanderings'. Testimonies like that of McLennan dovetailed with the timing of a committee ordered to investigate the sanitary state of the army in India. The report that the committee produced in 1863 was enormous (the final report ran to 1,902 pages over two volumes) and hugely influential. The production and release of this report was the final step towards the official re-introduction of the regulatory system in India, validating, as it did, the existing unofficial practices of surgeons and commanding officers in controlling and forcibly treating women deemed to be venereal threats.

On 31 May 1859, the Royal Commission on the Sanitary State of the Army in India, chaired by Lord Stanley, was appointed.[100] The Commission aimed to identify the habits of life and peculiarities of location believed to be at fault for illness among the soldiery across the subcontinent. In his task, Stanley was assisted by three prominent doctors: William Farr, an epidemiologist and medical statistician; James Ranald Martin, former president of the East India Company's Medical Board; and John Sutherland, a surgeon who had previously worked with Stanley on a 1858 Royal Commission on the Health of the British Army. The structure and concerns that guided the report were familiar and largely mirrored those of European surgeons serving in India such as Norman Chevers. Indeed, the Committee directly acknowledged the importance of his work, along with that of Thomas Graham Balfour and Alexander Tulloch.[101] As discussed in Chapter 4, Chevers, as co-editor of the *Indian Annals of Medical Science*, had published a series on the best means of preserving the health of European soldiers in India, which one of his contemporaries described as so important, that 'no medical officer should serve in India without studying it'.[102] Colonel Tulloch, whom the Committee directly questioned, was an early member of the Royal Statistical Society. He had entered the army as an ensign in 1826 and travelled to Burma, compiling tables and statistics of the health and mortality of the troops there.[103] The centrality of such studies to the Commission suggests that the work of European practitioners in India profoundly impacted the formulation of medical legislation both in England and across the Empire. This complicates the model that sees 'modern' medical practices and theories involved in a one-way transmission from Europe to the colonies. Here, the reverse was true, with surgeons and commanding officers returning from India to impart these theories and practices on the commission in London. The theory that supported the regulatory system, bound as it was with the insecurities of the military in India, was not 'created' in Europe and exported to India. While regulated prostitution operated in France and lock hospitals in London and Dublin, it was in India that these two methods were combined and refined to produce a more all-encompassing regulatory system.

The first section of the Commission's report is largely concerned with fatal diseases. It identified what it considered to be the most deadly ailments afflicting soldiers in India, namely: fevers, dysentery, liver diseases and epidemic cholera.[104] Of these, the report linked three very specifically to the geography of India. So, for example, dysentery, while not as frequent as fevers, was more fatal and was thought to manifest itself in the plains during the hot and rainy seasons.[105] Similarly, incidences of the acute inflammation of the liver were also believed to arise more often in the plains.[106] Cholera (the only one of the four not specifically situated in India), was nevertheless thought to be more fatal there than in England. While these fatal diseases were obviously of great concern to the army, they were difficult to prevent (especially when the vector itself was seen to be so intimately linked with the place). On the other hand, preventable diseases such as delirium tremens and syphilis, while not specific to India, 'enervated' the men and made them more vulnerable to other, more fatal, ailments. These 'fatal' and 'preventable' diseases cost the government an enormous amount of money, not just in treatment costs, but in the expense of soldiers rendered permanently invalid and sent back to Britain.

The Commission, unsurprisingly, focused much of their effort on understanding costs in India. To this end, the Report breaks down costs in some detail, explicitly linking the sanitary state of the army to the finances of India, noting that the former '... influences [finances] to a large extent.'[107] The dismal portrayal of the fatal nature of India and the subsequent cost to government was outlined in detail by a representation of four hypothetical armies, each composed of 33,615 19-year-old soldiers. Using the mortality statistics of soldiers in England as well as those for European civilians, military officers and soldiers in India, each army was 'measured' after 11 years of imagined service, which served to foreground the dramatic difference in fatality rates between those soldiers remaining in England and those sent to India.[108] In identifying the reasons behind these differences, the report blamed the soldier's lifestyle in India for his high death rates. The relatively good health of European civilians in India, the Report noted, was due to two key factors that counterbalanced any negative threats (such as weather) which might exist in India. First, the men were married and enjoyed the 'comforts of home', and second, they were not driven to 'drink or debauchery by ennui.'[109] This statement epitomises much of the Commission's findings on insalubrity. The language used to describe the men's health contains various coded references to domesticity or domestic life. The implication here is that 'home' is where the health is. Native troops, it was reasoned, were healthier as in most cases they had their families with them.[110]

Colonels and surgeons, when prompted, revealed their awareness of the informal controls put in place to control venereal disease already operational at numerous stations.[111] Colonel Swatman of the Bengal army assured the Committee that he was not certain why lock hospitals were initially discontinued, but speculated that it had something to do with economy. He assured the Committee that this loss was much lamented as they 'were always held to be a great check on the women.'[112] In March 1861, the Commission heard the testimony of Dr William MacLean, a former Deputy Inspector-General of Hospitals in India, now serving as Professor of Clinical and Military Medicine at Chatham Military College. MacLean told the Commission that '... the loss of efficiency from syphilis in India is now becoming quite a State question.'[113] He proclaimed that in one Queen's regiment, one woman had, over the course of two nights, 'utterly destroyed ten men'.[114] He went on to give further details of the number of men rendered unfit for duty by various malignant and aggressive forms of syphilis in India. Rather predictably, MacLean lamented the closure of the lock hospitals, and, when asked how they would work if re-opened, he suggested that were the police magistrates given the power to 'control' prostitutes, weekly examinations and treatments would be conducted with success.[115] MacLean declared that the establishment of lock hospitals in India would be more easily carried out in India than in Europe, due to the 'habits of the people' and the 'ready obedience with which they yield to authority'.[116] Again, defenders of the system were quick to use racial and social stereotypes in their justifications of the lock hospitals. Here, surgeons and officers capitalised on the supposed 'mildness' of character prevalent among Indians (which earlier observers insisted was due to factors which ranged from centuries of authoritarian rule to the oppressive climate to a vegetarian diet) to assure the committee of the success of such an endeavour. However, the timing of this assertion seems curious, coming as it did just after the 1857 rebellion. At a time when barbarism and brutality were watchwords running through most descriptions of India, MacLean's declaration suggests that while the uprising altered many colonial preconceptions about India, it was unable to shake the paradigm surgeons and commanding officers established around venereal disease. In fact, when it came to approaches to venereal disease control and drink, very little changed as a result of the Rebellion. If anything, in the minds of colonial administrators, the events of 1857 served to foreground the importance of a healthy body of European troops in maintaining imperial rule.

Medical returns from each of the stations supplemented testimonies like that of MacLean. The returns from the presidencies revealed that in 1860, out of every 1,000 men, venereal admissions stood at 345 in Bengal, 249 in Madras and 314 in Bombay.[117] After listening to various fearsome testimonies and reviewing returns regarding intemperance, morals and the slide towards venereal peril, the Commission recommended the '... re-organisation of the measures *formerly adopted in the three Presidencies*, with any improvements which subsequent experience and consideration may point out as being required to meet the necessities of each locality.'[118] The Commission's report did not represent any kind of real break with what surgeons in India and the medical profession generally had been arguing. What it did, however, was to organise these arguments and present them in a formal setting. Nor, as their recommendation made clear, was the proposed system to be 'imported' from Europe. The system formed and preserved by European surgeons in India was now given the green light to expand once again across the subcontinent. One Madras report drew this connection explicitly when it proudly boasted that their sanctioning the lock hospital system in 1859, had served to 'pave the way' for the measures now suggested by the Sanitary Commission.[119]

However, as the system moved towards broader acceptance, there were a few significant dissenting voices (largely from outside the military establishment). The lock hospitals opened in Karachi, Poona and Belgaum had failed in their objectives. Upon reviewing the evidence gathered on these lock hospitals which had been laid before him, the Governor of Bombay, Henry Bartle Frere, stated that he was '... not a believer in the efficacy of lock hospitals, nor of any exceptional measures.'[120] Nevertheless, he solicited the opinion of the Commander-in-Chief, William Mansfield. Mansfield, while agreeing with Bartle Frere that the hospitals had failed, framed the problem in a different manner. He asserted that the lock hospital experiment had failed as it was only a 'partial' one – that is, a hospital could be opened in a cantonment, but a neighbouring village would then 'remain unprotected against the inroads of disease.'[121] After reviewing the findings of Bartle Frere, Mansfield and the collected surgeons who gathered evidence, the Bombay Military Department agreed that the cost of the lock hospitals was simply not worth the paltry results they produced and suggested their abolition.

Regional differences between the Presidencies intensified once again. As Bombay under Bartle Frere was turning away from the lock hospital system (once again, tellingly, on account of cost), Bengal was once again moving towards it. Act XXII of 1864 or, An Act to Make Provision for the Administration of Military Cantonments, received the assent of the Viceroy on 1 April 1864 and was all-encompassing in its goals. Initially, the Cantonment Act (as it was known) was only applicable to Bengal, but it contained a provision to enable its extension to any area under the administration of the Government of India as well as any place, even those not under direct British control where British troops were cantoned. The Cantonment Act not only targeted women deemed to be a threat to military health, but more broadly sought to regulate the '... administration of Civil and Criminal Justice and the superintendence of Police and Conservancy, for protecting the public health within the limits of Military Cantonments'.[122]

In line with previous acts to regulate cantonment land, the 1864 Act outlined restrictions and rules regarding land and the occupation of property in the cantonment. In addition, Section XIX of the Act provided not only a definition and prohibition of 'public nuisances', but laid out the rules for the general maintenance of health and sanitation within the cantonment. It specifically allowed for the inspection and control of 'houses of ill fame' for the purposes of controlling venereal disease.[123] Moreover, Section XXV of the Act allowed for its extension (whenever necessary) *beyond* cantonment limits to preserve the health of the troops. The Cantonment Act wove together the various threats seen to be stalking the cantonments – venereal disease, intemperance and dirt – and, unsurprisingly, drew links between the three. The military quickly implemented the Act across Bengal, but, as lock hospitals were already operational in Madras, it was rolled out unevenly both there and in Bombay.

Despite the aims of the Royal Commission and those drafting the 1864 Act, like its predecessors, the Act operated erratically across India. However, the advocates of the system, perhaps gaining confidence from the Cantonment Act and the statutory British version of the Act (which came in the form of the 1864 and 1867 Contagious Diseases Acts), pushed further once again. Advocates of regulation once more attempted to fill the gaps in the system's coverage, seeking a more formal and punitive system, which could be established across the subcontinent.

Doctor Fabre-Tonnerre, Calcutta's first Health Officer, was one of the driving forces behind the introduction of the Indian Contagious Diseases Act of 1868 (Act XIV).[124] To this end, Fabre-Tonnerre spat out a vitriolic attack against Indian 'prostitutes', arguing that it was the government's moral duty to introduce a series of regulations to curtail the degradation and threat the women represented. He noted that

> Nowhere are the lower classes of the Christians and Natives more debased, dissolute, and unclean than in the metropolis of India. Everything amongst them bears the stamp of a *total absence of every moral, religious and social feeling*. Hence the great number of prostitutes, who not only *swarm* in the bye-lanes and back-slums of Calcutta, but who *infest* our principal thoroughfares, polluting the atmosphere of our neighbourhood, and who, by their indecent conduct, scandalize the morals of the population in the midst of which they are permitted to live.[125]

He continued on with this outraged sermon, noting that it was the *duty* of municipal bodies and government to remove such 'nuisances'. This language of 'infestation' and immorality suggests a pestilential epidemic instead of a group of poor, working women. Fabre-Tonnerre sought to stress, in no uncertain terms, that prostitutes constituted a wholly separate class from the general public. Their constant violation of both moral and divine laws rendered them, in his view, 'liable to special legislation which, it may be alleged ... interferes with the liberty of the subject.'[126] Now, more than ever, he argued, society needed to protect its interests against such threats. He drafted what he titled 'An Act for the Prevention of Contagious Diseases and the Control of Prostitution in Calcutta,' and forwarded this to Calcutta's Chief of Police.

Government and military officials rapidly escalated Fabre-Tonnerre's demands and 1868 saw the passage of the Indian Contagious Diseases Act. The Act could be applied *anywhere* in India where it was deemed necessary to do so and this would soon include not only cantonments, but the presidency cities as well. While the Indian Contagious Diseases Act varied little from previous systems of regulation, it formalised earlier procedures, demanding (and generating) mountains of paperwork in its wake. Medical and military officers generated protocols for every aspect of the Act, from the proper wording used to notify the public of an extension of the Act to an area, to the forms completed on each prostitute under its remit (see Figure 5.1 for an example of the latter). Government ordered regular, comprehensive reports on the functioning of the lock hospitals. These detailed every aspect of their operation, from the diet provided to patients (see Table 5.5) to the floor plans for specific sites (see Figure 5.2). The Act also went further in its willingness to use force against women who failed to obey its stated terms. In Calcutta alone, an average of 12 women were arrested daily between the 1870s and 1888 for breach of the rules.[127]

Methodical record-keeping and detailed diet plans were not enough to save the system from the fate of its predecessors; it, like them, was unsuccessful. Once again, the expense of the system proved the biggest bone of contention. Disagreements on funding the system soon broke out between municipalities and the military. While Government had agreed to pay a proportion of the Act's operating costs, as it was asserted that such measures benefitted the 'public', municipalities were ordered to pay

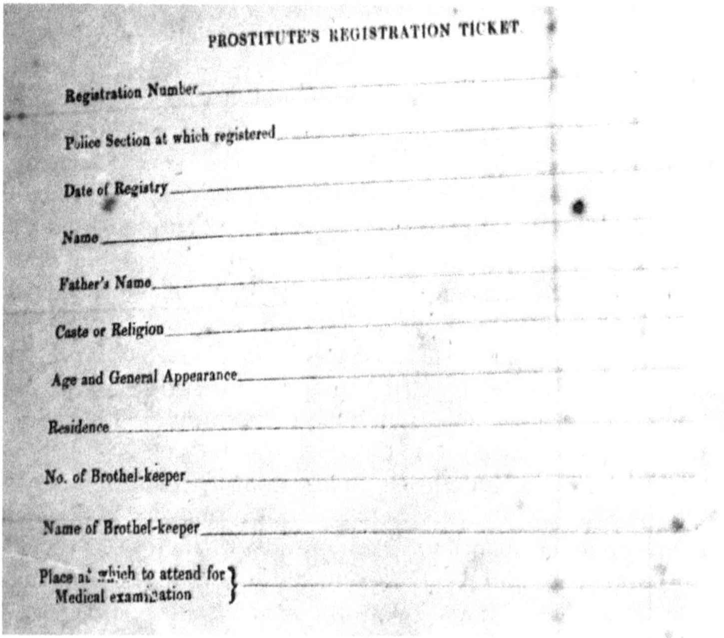

PROSTITUTE'S REGISTRATION TICKET

Registration Number________

Police Section at which registered________

Date of Registry________

Name________

Father's Name________

Caste or Religion________

Age and General Appearance________

Residence________

No. of Brothel-keeper________

Name of Brothel-keeper________

Place at which to attend for Medical examination________

Figure 5.1 Prostitute's registration ticket
Source: *Prostitute's Registration Ticket*. WBSA, General (Sanitation). Proceedings, No 67, April 1869.

Table 5.5 Diet tables for patients in lock hospitals

	A – Soup diet
Carbon 2038.31	Nitrogen 114.54
Breakfast	2 oz rice and 10 oz of milk with ½ oz sugar to make congee
Dinner	1 pint mutton and chicken broth, 2 oz rice
Supper	2 oz sago, 10 oz milk and ½ of sugar
	B – Congee diet
Carbon 1115.19	Nitrogen 38.22
Breakfast	1 oz rice and 4 oz milk to make congee
Dinner	2 oz sago and ½ oz sugar to make congee
Supper	1 oz rice and 4 oz milk to make congee
	C – Milk diet*
Carbon 2988.43	Nitrogen 175.06
Breakfast	2 oz rolong [sic]**, 10 oz milk and ½ oz sugar to make congee
Dinner	6 oz rice and 10 oz milk and ½ oz sugar to make rice milk
Supper	2 oz rolong and 10 oz of milk and ½ oz sugar to make congee
	D – Ordinary diet A
Carbon 3399.41	Nitrogen 196.88
Breakfast	8 oz bread, 1 pint of tea, 2 oz milk and ½ oz sugar, butter ½ oz
Dinner	6 oz mutton, ½ chicken or 1/3 foul, 6 oz potatoes, 4 oz rice with curry stuff ½ oz
Supper	4 oz bread, 1 pint of tea, 2 oz milk and ½ oz sugar
	E – Ordinary diet B
Carbon 3428.48	Nitrogen 200.92
Breakfast	2 oz rice to make congee, milk 4 oz
Dinner	Rice 4 oz, dholl 3 oz, ghee ½ oz, to make kedgeree, 4 oz vegetables, with ½ oz curry stuff
Supper	8 oz bajree or wheaten bread, 6 oz mutton, or ½ chicken with ½ oz curry stuff
	F – Ordinary diet C
Carbon 3922.70	Nitrogen 194.76
Breakfast	2 oz rice to make congee, milk 4 oz
Dinner	4 oz rice, 3 oz dholl, ½ oz ghee to make kedgeree, 4 oz vegetables with ½ oz curry stuff
Supper	8 oz wheaten bread or bajree, 4 oz milk and ½ oz sugar

* The presence of a 'milk' diet suggests that officials were attempting to make concessions for Brahmin women, such as the Kulin widows who it was felt, were pushed into prostitution.
** Hobson-Jobson describes 'rolong' as a fine flour.
Source: Diet Tables for Patients in Lock Hospitals. November 1877, MSA, Military Department (1875–1878), Vol. 2366, 217.

GROUND PLAN
OF
LOCK-HOSPITAL, BAREILLY.

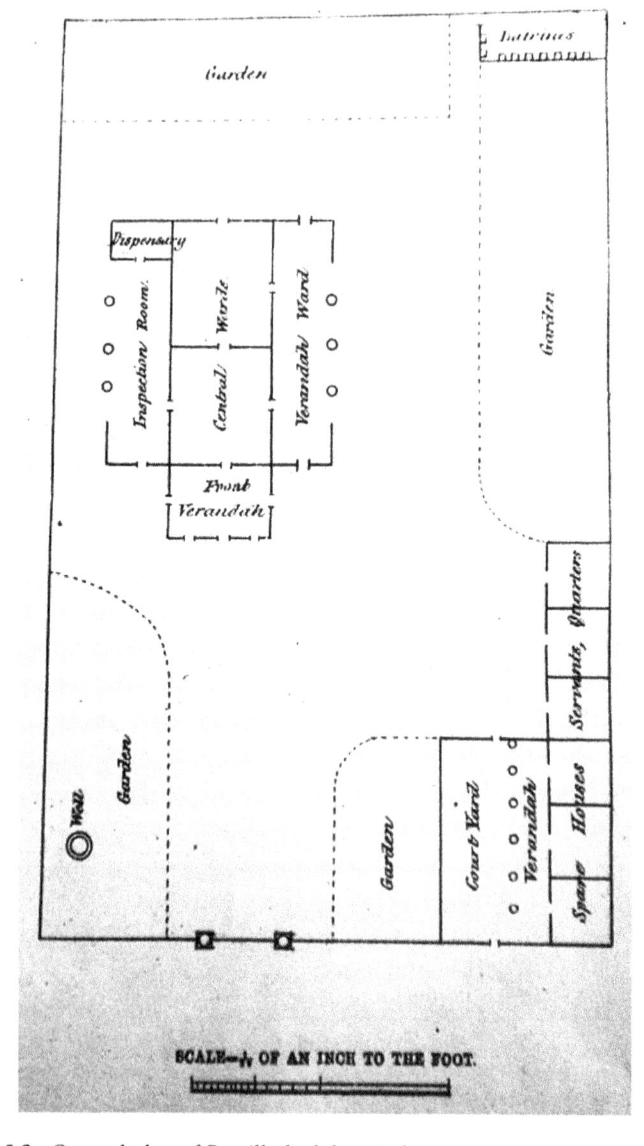

Figure 5.2 Ground plan of Bareilly lock hospital
Source: *Ground Plan of Lock-Hospital, Bareilly. Report of the Lock Hospitals of the North West Province during the year 1875*. NAI, Home (Sanitary). Proceedings, July to December 1875.

the remainder. Municipal governments and boards (no doubt with limited budgets) resented this imposition and rightly noted that the Act was one which almost entirely benefitted the military. These disagreements came to a head in the late 1870s in Bombay, when the municipal government withdrew its funding for the Act and effectively halted the system. While military fears had elevated the lock hospital system to critical strategic importance, the municipalities were unfazed and unconvinced by such perceived 'threats'. Once again, the decision to abolish the lock hospitals rested on cost. Viceroy Ripon suggested its abolition in 1882, noting that the money used for lock hospitals could be much better spent on implementing a system of public sanitary works.[128] Accordingly, the system was part-suspended in 1883. However, it continued to operate in some places until 1888, when the House of Commons forced the closure of remaining lock hospitals and the full repeal of the Act.

Conclusion

The arguments of those demanding the re-introduction of lock hospitals from the 1830s were no more sophisticated or rational than those made in the early nineteenth century. In fact, they had changed little, if at all. The concerns in the late eighteenth century which prompted the early experiments to stem rising venereal disease admissions, solidified to form a largely unshakeable theory that remained the dominant model for venereal disease control throughout the nineteenth century. The operation of the lock hospital system illustrates the ways in which the policies of the colonial state were undermined – often by unexpected actors. The renegade policies of surgeons and commanding officers, either quietly or openly, through their steadfast devotion to the lock hospital system (whether it was authorised or not), subverted the state's authority.

By the mid-nineteenth century, the mentality that supported the lock hospital system pervaded medical and military thinking and definitively shaped the construction and formalisation of 'public health' in India. The framework established by the lock hospital system that emphasised bodily control was applied to disease control more generally. The insecurities and fears that shaped and determined the formation and operation of the system never went away. Supporters of regulation remained convinced that the system was essential to prevent the total disintegration of the army through venereal disease. In this context, we can more clearly understand the reasons behind the heated debates that raged in the 1890s, forcing the British government to the edge of a 'constitutional crisis'.

Conclusion

The European soldier in India presents a curious figure – a contradiction at the core of empire. He was at once reviled for his supposed degeneracy (thought to be a natural extension of his class) and relied upon to give support to racial structures within the military, itself the beating heart of Crown and Company rule. As such a valuable, but potentially unstable asset, considerations regarding his health were central to Company (and later Crown) decision-making. The East India Company was first and foremost a commercial enterprise whose main concerns were profitability and security. It followed that the Company's army was to be economically and carefully maintained; allowing the Company not just to remain afloat in a hostile sea of competitors, but aggressively to expand. However, the health problems of the troops, and more specifically, the European soldiery, persistently upset this ideal. Dysentery, cholera and fevers fatally struck down scores of men, while at any given time in the nineteenth century, venereal disease and intemperance claimed roughly one-quarter to one-third of hospital admissions. Given the Company's anxieties about the stability of their position in India, the spectres of venereal disease and intemperance took on added significance.

Commanding officers and surgeons rolled out the lock hospital system and cantonment regulations with the primary intention of preserving the health of the European rank-and-file. Fiscal constraints were only part of the calculation that determined the shape of the system; these were complemented by the view that saw European soldiers as violent and volatile. This combination led to the cautious construction of policies to cater for their perceived needs. Ideas about the men's class and character led officers to believe that any enforced or prolonged treatment would provoke tremors of discontent which would lead to open mutiny. This book has demonstrated the significant ways in which these regulatory regimes reflected the broader anxieties of empire. It has shown that the implications of these calculations extended far beyond the barracks and hospital and reached into the intimate lives of European soldiers and officers and the legions of Indians whose lives involved, in any way, interaction with the army.

When it came to the health of the European troops, commanding officers and surgeons pushed their own demands and agendas instead of conforming to the dictates emerging from the Writer's Building or Government House in Calcutta and in so doing, often directly contravened 'civil' law. Thus, while it has been argued that Indian collaborators undermined the colonial state, in the case of the attempted control of venereal disease, the Company's British servants carried out such subversion. This work demonstrates how these breaks in the line of control went further still. The fear of the European soldiery that shaped the lock hospital system highlights the precarious control of officers over even (or, perhaps more accurately, *especially*) their European troops. In order to understand the Company state, we must first understand the perception it maintained of European soldiers. The focus on the mutiny amongst the sepoys in 1857 has led to retrospective assumptions about loyalty within the armies. However, for much of the first half of the nineteenth century, it was not mutiny amongst the sepoys, but one among the European troops which officers believed the greater possibility.

The lock hospital system also grew out of a desire to spare the European troops the strain of lengthy periods of treatment; these were physically and fiscally debilitating for the army. In the case of venereal disease, not only could Indian bazaar 'prostitutes' be treated at a fraction of the cost of the European men, but it was assumed that as a group, the women would be easier to control and any 'mutinous' behaviour on their part could be swiftly put down, without as great a fear of the repercussions. And yet, the women targeted by the system *did* rebel, perhaps not with the force of arms, but with tactics of everyday resistance, most commonly with evasion or concealment. The system was designed so that these women would absorb the brunt of the treatments – many of which remained dangerous. Its supporters asserted that if successful, the lock hospital system would provide a 'healthy', regulated prostitute population in the cantonment to safely serve the men's sexual needs. Yet, aware of the invasiveness of such a system and the potential upset it might provoke, medical and military officers quickly sought to establish a medical and moral cordon around the women targeted by it. This targeting forced certain women into the category of 'prostitute' who had never previously been considered as such. This was reinforced by European value judgements that situated the women in the 'criminal' category and placed them outside, and beyond the pale of 'respectable' Indian society. As such, the lock hospital system not only had a direct physical impact on the women themselves, but broader ramifications for society and empire.

The notion that venereal disease was gendered – originating with 'dangerous' women – both informed and reinforced the lock hospital system. This was certainly not a new idea, but surgeons and commanding officers were committed to the view that bazaar women rather than soldiers, were the vectors of infection. Not only were women condemned for transmitting disease, but officers held them responsible for the failure of the system to significantly reduce hospital admissions among the men. Officers dismissed or simply ignored the fact that such resistance might be a perfectly understandable reaction to an invasive, deeply unpleasant system that deprived women of their livelihoods and separated them from their kin and dependants.[1] Instead, the women's avoidance of the system was described as irrational, and blamed on Indian 'avarice', 'craft' or 'superstition'. This contributed to the development of a new discourse on the supposed racial, social and moral differences between European and 'colonial' subjects.

In their attempts to garner support for the regulatory system, surgeons and commanding officers not only targeted and demonised the Indian 'prostitute', but re-wrote the very category. 'Prostitutes' became not just a 'social evil', but a very real physical danger. The alliance of military and medical men who worked throughout the nineteenth century to protect the European soldiery from all manner of health risks, in the calculating language of 'modern', 'rational' medicine, portrayed anyone or anything that stood in their way as threats. As we have seen, this increasingly included not only the women affected by the system, but a wide assortment of Indians living in and around the cantonments. This book has highlighted these 'intermediaries' of empire to help us understand how they were transmuted by officials into a collective threat which in turn transformed both the colonial state and European understandings of India.

This movement among surgeons was representative of a larger process of shifting the definitions of 'tradition' and mores in colonial India. Fear dictated not just the colonial state's policy on venereal disease control, but more broadly guided attitudes towards the military control of land. What this study brings to the fore is the 'lowest common denominator' approach which saw the colonial state compress diverse groups into base categories, whether they were Indian women or European soldiers. As individuals became vague stereotypes, the same fear that motivated earlier Company policies intensified, perhaps, we might suggest, due to the fact that these groups very often little reflected reality. Medical and commanding officers pointed to these supposed racial, social and moral distinctions in their calls for greater military control over land and occupants. These demands pushed out the physical and social boundaries of military control in India. As the century progressed, army surgeons and commanding officers implemented an increasingly penal regime in and around cantonments where European troops were stationed. These regulations reinforced the idea that European health in India could only be achieved through managed segregation.

The regulatory framework was not simply imported from Europe and enacted in India. Instead, it grew out of a complex set of circumstances and conditions. The operation of the lock hospitals over the course of the nineteenth century points to the myriad ways in which ideas and practices on everything from medicines to human difference were *exported* from India to Europe. The experience of venereal disease control directly contradicts any argument which projects colonial policies as simply one-way. The reality was much more complex and nuanced. Not only did European surgeons borrow heavily from 'native' practices and treatments, but the system relied on Indian intermediaries to operate. Medical boards and individual surgeons continued to absorb indigenous ideas of disease theory and treatment, at times unconsciously slipping them into their own theories. Based on their experiences in India, European surgeons portrayed their contributions to the medical field as distinct and unique. What is more, they were confident that their knowledge of Indian subjects could contribute to broader understandings of human difference.

The lock hospital system at no point succeeded in its goals. Yet, despite its repeated failures, surgeons clung to the system as the only hope in the battle against venereal disease in India. Why was this the case? In part this was linked to the very founding of the lock hospitals themselves. In the absence of any real cure, any reduction in the number of men needing to undergo unpleasant treatment and extended stays in hospitals was welcomed. As the lock hospital system faced abolition in the 1830s, the medical boards, societies and journals reinforced surgeons' demands and gave them a more powerful voice. Through these networks, medical men developed a 'common sense' on the best means to control venereal disease. Their unitary voice was unfailingly constant in its support for the lock hospital system. Through these media, surgeons and commanding officers found a much more effective means to lobby the colonial state.

In contrast to earlier works on the colonial state's attempt to control venereal disease, this book has shown that the lock hospital system was not simply a medical operation, but a self-replicating construct. At its heart was the idea that undesirable, 'polluting' elements of society could be managed and controlled for the greater good. More critically, perhaps, these systems of bodily regulation reinforced the idea that there was a sure link between 'moral' control and the maintenance of colonial rule. In trying to protect the health of the European soldiery and order the cantonment, military and medical men ensured that a medically-driven 'morality' became the new 'norm'. In their approach to venereal disease among the European troops, military and medical officers figuratively transformed the disease – making it into a much more significant imperial threat. As soon as surgeons and commanding officers began to blame the system's failure on its targets and Indian collaborators, the explanations arising stressed racial difference and presumed deficiency. The regimen reinforced the colonising mentality, which used Indian inferiority to justify the continuance of empire. What is more, the system contributed to the European construction of the Indian 'character'. Patterns of 'prostitution' and morally 'questionable' (to European eyes) traditions in India reassured Europeans of their position in the global hierarchy of peoples and emphasised the importance of their 'civilising mission'.

This book returns the military to the central position that it occupied during the Company raj. Given the paramountcy of the military to both Company and Crown raj, studies exploring the internal dynamics of the army have been relatively thin on the ground. Fortunately, this has begun to change in recent years, with the emergence of a 'new military history' of South Asia. It is hoped that this work will contribute to this debate, not simply as a social history of the army itself (and, by extension, the colonial state), but as a study in the broader impact of the army on Indian society. This book has demonstrated how military priorities infiltrated every aspect of Company and Crown regimes to a hitherto unrecognised degree – from the judiciary to routine, daily interactions with Indian subjects. Douglas Peers has shown us that the colonial establishment in India was, what he deems a 'Garrison State'.[2] This study takes Peers' framework further, to show how a military-led mindset crept into the everyday – where seemingly unrelated aspects of life in colonial India were shaped by the demands of the military. However, perhaps more importantly, it has shown that this narrative was complicated by a contradiction at the very heart of colonial rule. This *was* a garrison state, but one which was riddled with minor mutinies and disorderly behaviour from its very gate-keepers. Far from being a monolith, the Company was frequently undermined by its own servants.

The reforms and regulations surrounding the issue of 'vice' highlight the inconsistencies and vulnerabilities of the Company state. Born and sustained by this complex set of fears, the lock hospital system remained the dominant paradigm for the treatment of venereal disease until the early twentieth century. Like the vices it sought to control, the regulatory system proved a many-headed hydra; once unleashed, it took on a life of its own and became difficult to contain. The machinations of military and medical officers, aimed at safeguarding the health of the soldiery, quickly travelled beyond military space to profoundly alter colonial state and society.

Appendix 1: [Extract] Queries respecting the human race, physical characters[1]

1. State the general stature of the people, and confirm this by some actual measurements. Measurement may be applied to absolute height, and also to proportions, to be referred to in subsequent queries. The weight of individuals, when ascertainable, and extreme cases, as well as the average, will be interesting. What may be the relative difference in stature and dimensions, between males and females?
2. Is there any prevailing disproportion between different parts of the body? As, for example, in the size of the head, the deficient or excessive development of upper or lower extremities?
3. What is the prevailing complexion? This should be accurately defined, if possible by illustrative and intelligent example, such as by comparison with those whose colour is well known. The colour of the hair should be stated, and its character, whether fine or course, straight, curled, or woolly. The colour and character of the eyes should likewise be described. Is there, independently of want of cleanliness, any perceptible peculiarity of odour?
4. The head is so important as distinctive of race, that particular attention must be paid to it: is it round or elongated in either direction, and what is the shape of the face, broad, oval, lozenge-shaped, or of any other marked form? It will contribute to facilitate the understanding of other descriptions, to have sketches of several typical specimens. A profile, and also a front view should be given. In the profile, particularly notice the height and angle of the forehead, the situation of the meatus auditorius, and the form of the posterior part of the head. It will also be desirable to depict the external ear, so as to convey the form and proportion of its several parts. The form of the head may be minutely and accurately described by employing the divisions and terms introduced by craniologists, and the corresponding development of moral and intellectual character should in conjunction be faithfully stated. So much of the neck should be given with the profile as to show the setting on of the head. The advance or recession of the chin, and the character of the lips and nose, may likewise be given in profile. The front view should exhibit the width of forehead, temples, and cheek-bones, the direction of the eyes, and the width between them: the dimensions of the mouth. When skulls can be collected or examined, it would be desirable to give a view in another direction, which may even be done, though with less accuracy, from the living subject. It should be taken by looking down upon the head from above, so as to give an idea of the contour of the forehead, and the width of the skull across from one parietal protuberance to the other.
5. State whether the bones of the skull are thick, thin, heavy, or light. Is it common to find the frontal bone divided by a middle suture or not?

[...]

11. Where a district obviously possess two or more varieties of the human race, note the typical characters of each in their most distinct form, and indicate to what known groups or families they may belong: give some idea of the proportion of each, and state the result of their intermixture on physical and moral character. When it can be ascertained, state how long intermixture has existed, and of which the physical characters tend to predominate. It is to be observed, that this question does not so much refer to the numerical strength or political ascendancy of any of the types, but to the greater or less physical resemblance which the offspring may bear to the parents, and what are the characters which they may appear to derive from each: whether there is a marked difference arising from the father or the mother belonging to one of the types in preference to another; also whether the mixed form resulting from such intermarriage is known to possess a permanent character, or after a certain number of generations to incline to one or other of its component types.
12. Any observation connected with these intermarriages, relating to health, longevity, physical and intellectual character, will be particularly interesting, as bringing light on a field hitherto but little systematically investigated. Even when the people appear to be nearly or quite free from intermixture, their habits, in respect of intermarriage within larger or smaller circles, and the corresponding physical characters of the people, will be very interesting.

    [...]

32. Is chastity cultivated, or is it remarkably defective, and are there any classes amongst the people of either sex by whom it is remarkably cultivated, or the reverse, either generally or on particular occasions.
33. Are there any superstitions connected with this subject?
34. What are the ceremonies and practices connected with marriage?
35. Is polygamy permitted and practised, and to what extent?
36. Is divorce tolerated, or frequent?
37. How are widows treated?

    [...]

57. Is there any division of clans or castes?
58. What are the privileges enjoyed or withheld from these?
59. What care is taken to keep them distinct, and with what effect on the physical and moral character of each?

    [...]

77. Are the people addicted to religious observances, or generally regardless of them?

Appendix 2: Queries relative to the nature and treatment of the venereal disease in India

With mercury	Without mercury
1. Will have the goodness to state generally the extent of your experience in the treatment of venereal ulcerations on the penis with and without bubo during your residence in India.	The same question without mercury.
2. Describe the general character of the sores so treated in the most accurate terms.	Ditto.
3. Of those treated by you how many on average have had secondary symptoms.	Ditto.
3/2 When secondary symptoms have occurred, have they been of a more obstinate character after the mercurial or non mercurial mode of treatment.	Ditto.
4. Describe the kind and characters of the secondary symptoms and the average period which intervened between their appearance and the treatment for the primary sores.	Ditto.
5. Describe the treatment both general, and local, usually employed by you for the cure of primary venereal cases.	Ditto.
6. The same question for secondary symptoms.	Ditto.
7. State the average periods occupied in the cure of primary symptoms distinguishing with and without Bubo.	Ditto.
8. Were the subjects of all such cases fit for immediate exertion or exposure to the inclemencies of the weather, such as might be expected in the course of active military duty.	Ditto.
9. State the average period occupied in the cure of secondary symptoms.	Ditto.
10. Have you observed that the mode of treatment pursued exposed the subjects of it, to future attacks of disease and of what particular nature.	Ditto.

(*continued*)

Appendix 2 Continued

With mercury	Without mercury
11. Were any of the subjects of the cases treated by you and how many so injuriously affected by the disease that had they belonged to the Military profession, they must have been invalided.	Ditto.
12. State the present condition of the subjects of the cases treated by you, for primary and secondary symptoms, noting any diseases to which they have been subject during the interval.	Ditto.
13. What is the comparative number of secondary symptoms after each mode of treatment.	Ditto.
14. Have you observed any causes which in your opinion may lead to the appearance of secondary symptoms, either in the season, mode of life, constitution or manner in which the primary symptoms have been treated.	Ditto.
15. What number of cases of primary, or secondary symptoms have occurred, distinguishing each, for the cure of which, it has been deemed absolutely necessary to have recourse to mercury, from the non mercurial mode of treatment having appeared to fail after a fair trial.	
16. What has been the average period during which in the cases alluded to, in the last question, the non mercurial mode of treatment was tried, and what were the remedies and dressings employed.	
17. State the symptoms which in your opinion imperiously called for the Mercurial in those cases.	
18. Have those patients, who have not used mercury, and who have been affected with secondary symptoms, ever had a recurrence of their symptoms, after having recovered from them without Mercury, and where mercury has been used.	
19. When scabby eruptions have been observed is it known whether the scabbing existed from the commencement of the eruption, or if it only appeared on its decline.	
20. Is an alternative course of mercury sufficient for the cure of this eruption in an incipient state, and in its decline.	
21. Do large tumours, or those swellings termed gummata ever occur where Mercury has not been employed.	
22. In how many cases has caries of the bones been observed after each mode of treatment.	
23. Have the goodness to state the number of cases of nodes which have succeeded the treatment with mercury and without mercury.	

(*continued*)

Appendix 2 Continued

With mercury	Without mercury

24. What are the characters of the primary ulcer and of the secondary eruptions in those cases in which nodes occur where Mercury has not been employed.
25. Have the goodness to state the number of cases of iritis which have succeeded the treatment with mercury and without mercury.
26. Although it is believed, that iritis occurs more frequently after the non mercurial mode of treatment, and that it is most advantageously combated by mercury, your opinions on the point is requested.
27. When more than one person has had intercourse with the same diseased woman, are all the resulting affections similar, if otherwise, what differences are observed in primary and secondary symptoms.
28. Should particular primary symptoms be found to give rise to particular secondary ones what are the characters of the sores of each.
29. Are you of opinion, that the character of the chancre depends on the virus, on the constitution or habit of body, or do you consider that it may be made to assume different characters according to the treatment.
30. The same question regarding secondary symptoms.
31. Is it observed that Gonorrhoea ever gives rise to secondary symptoms, of any kind, if so describe them.
32. Is there any other variety observable in the persons contracted by promiscuous intercourse, than that which produced Gonorrhoea and chancre.
33. Have you observed any cases in this country, resembling the framboesia or yaws of the West Indies.
34. If so are you aware of any peculiarity in the primary symptoms, which occurred in such cases.
35. Do you consider the venereal disease, which prevails in this country more or less severe than that which prevails in Britain.
36. With reference to the constitution, mode of life, and diseases which prevail among the natives in the different provinces of India, do you consider them more or less fitted for the mercurial mode of treatment than the natives of our own country.
37. In Natives, who have lived very poorly, have you observed primary ulcers to assume a phagedenic character, and to be quickly succeeded by numerous extensive superficial secondary ulcerations, over the whole body, where mercury has been too freely or too rapidly administered.
38. Has mercurial erythema been frequent occurrence in your practice, and has it ever proved fatal and how often.
39. Are you of opinion that nodes and caries of the bones of the nose, are generally benefited by the administration of mercury or otherwise.
40. Have you observed any one Mercurial to be more relied on as an anti-syphilitic than the others and what particular preparation.

(*continued*)

Appendix 2 Continued

With mercury	Without mercury

41. Have you observed any one mode of administering mercury more useful than another in speedily inducing a benignant ptyalism.
42. Have you had recourse to fumigation with mercury. What preparation have you used, in what dose, and to what extent both generally and locally.
43. If you possess authentic information, will you state the general practice in use among the Natives in such cases and the success with which it is attended.
44. If you possess authentic information will you detail the various modes in which the natives administer mercury.
45. Are you aware of the local applications to which the natives have recourse in venereal ulcerations, and with what success.
46. Will you have the goodness to state any information consistent with the object of these enquiries and not embraced by them which in your opinion would be interesting or useful.'

Source: Circular Letter to the Superintending Surgeons of Divisions, 14 October 1831. NAI, Home (Medical). Bengal Medical Board Proceedings, 17 October 1831.

Notes

Introduction

1. William Combe, *The Grand Master; or, Adventures of Qui Hi? in Hindostan. A Hudibrastic Poem in Eight Cantos. Illustrated with Engravings by Thomas Rowlandson* (London: Thomas Tegg, 1816), 47–8. Italics in original. 'Qui Hi', was a nickname for Bengal Anglo-Indians and more specifically, those serving in the Bengal army. It is derived from the phrase used to call servants meaning 'Is anyone there?' and was customarily called out when a Bengal officer entered the mess.
2. Rudyard Kipling, *Barrack-Room Ballads and Other Verses* (London: Methuen & Co., 1892). Of course, while the fictional 'Qui Hi' is an officer cadet, the descriptions that swirl around him throughout the poem were those more often affixed to the rank-and-file.
3. An interesting comparison could be made between the British vision of its soldiery with that of, say, France or Tsarist Russia. Hew Strachan makes a similar point with regard to patterns of recruitment for the army in Britain in his discussion of the army reforms of the 1830s to 1850s. See Hew Strachan, *Wellington's Legacy: The Reform of the British Army, 1830–1854* (Manchester: Manchester University Press, 1984), 50–1.
4. David Omissi has made a similar argument with regard to sepoy troops, noting for example, that the low level of sepoy illiteracy was the result of policy, not chance. See David Omissi, *The Sepoy and the Raj: The Indian Army, 1860–1940* (London: Macmillan Press, 1994), 27.
5. For much of the century, there was marked resistance to any proposals for diversions such as libraries or theatres for the men as there was to such schemes as savings banks for soldiers. In the early nineteenth century, men enlisted to the army for 21 years. This lengthy term, combined with the weakened health of many (for some, even before they left Britain) meant that most died in India. The terms of service were changed by the 1847 Time of Service in the Army Act, which reduced the term to 10 years. By this time, morbidity had also decreased considerably. Cardwell's 1870s reforms reduced the years of service even further.
6. John Mercier Macmullen, *Camp and Barrack Room; or, The British Army as It Is, by a Late Staff Sergeant of the 13th Light Infantry* (London: Chapman and Hall, 1846), 75.
7. National Archives of India (hereafter NAI) Letter to Warren Hastings from Charles Smith, Commanding in Madras. 16 April 1781. NAI Secret Consultations, 27 April 1781, no. 1.
8. This model was proposed by George Basalla, 'The Spread of Western Science', *Science* 156, 3775 (1967).
9. David Arnold, *Science, Technology, and Medicine in Colonial India* (Cambridge: Cambridge University Press, 2000), 13.

10. England saw a series of Acts which were modified three times over the period from 1864 to 1886. For a detailed analysis of the Contagious Diseases Acts in England, see Judith Walkowitz, *Prostitution and Victorian Society: Women, Class and the State* (Cambridge: Cambridge University Press, 1980); Philip Howell, *Geographies of Regulation: Policing Prostitution in Nineteenth-Century Britain and the Empire* (Cambridge: Cambridge University Press, 2009); E. M. Sigsworth and T. J. Wyke, 'A Study of Victorian Prostitution and Venereal Disease', *Suffer and Be Still*, 77–99 ed. Martha Vicinus (Bloomington: Indiana University Press, 1972); Philippa Levine, 'Public Health, Venereal Disease and Colonial Medicine in the Later Nineteenth Century', *Sin, Sex and Suffering: Venereal Disease and European Society Since 1870*, 160–72 eds. Roger Davidson and Lesley Hall (London: Routledge, 2001).
11. Philippa Levine, *Prostitution, Race, and Politics: Policing Venereal Disease in the British Empire* (New York: Routledge, 2003); Myna Trustram, 'Distasteful and Derogatory? Examining Victorian Soldiers for Venereal Disease', *The Sexual Dynamics of History: Men's Power, Women''s Resistance*, 154–64 ed. London Feminist History Group (London: The Pluto Press, 1983); Kenneth Ballhatchet, *Race, Sex and Class Under the Raj: Imperial Attitudes and Policies and Their Critics, 1793–1905* (London: Weidenfeld and Nicholson, 1980). For a study of venereal disease control in the early nineteenth century, see Douglas Peers, 'Soldiers, Surgeons and the Campaigns to Combat Sexually Transmitted Diseases in Colonial India, 1805–1860', *Medical History* 42, 2 (1998). Douglas Peers also explores other forms of vice, such as intemperance, see Douglas Peers, 'Imperial Vice: Sex, Drink and the Health of the British Troops in North Indian Cantonments, 1800–1858', *Guardians of Empire: The Armed Forces of the Colonial Powers c. 1700–1964*, 25–52 eds. David Killingray and David Omissi (Manchester: Manchester University Press, 1999).
12. For the Colonial Acts, see Levine, *Prostitution, Race, and Politics*; Elizabeth van Heyningen, 'The Social Evil in the Cape Colony 1868–1902: Prostitution and the Contagious Diseases Acts', *Journal of Southern African Studies* 10, 2 (1984); Howell, *Geographies of Regulation: Policing Prostitution in Nineteenth-Century Britain and the Empire* (Cambridge: Cambridge University Press); Ashwini Tambe, *Codes of Misconduct: Regulating Prostitution in Late Colonial Bombay* (Minneapolis, MN: University of Minnesota Press, 2009).
13. Philippa Levine, 'Venereal Disease, Prostitution and the Politics of Empire: The Case of British India', *Journal of the History of Sexuality* 4, 4 (1994), 580.
14. Mark Harrison, *Public Health in British India: Anglo-Indian Preventive Medicine, 1859–1914* (Cambridge: Cambridge University Press, 1994), 73.
15. Sumanta Banerjee, *Dangerous Outcast: The Prostitute in Nineteenth Century Bengal* (New York: Monthly Review Press, 1998), 153.
16. Sarah Hodges, '"Looting" the Lock Hospital in Colonial Madras during the Famine Years of the 1870s', *Social History of Medicine* 18, 3 (2005), 390.
17. Antoinette M. Burton, *Burdens of History: British Feminists, Indian Women, and Imperial Culture, 1865–1915* (Chapel Hill; London: University of North Carolina, 1994).
18. Levine, *Prostitution, Race, and Politics*.
19. Raymond Callahan, *The East India Company and Army Reform, 1783–1798* (Cambridge, MA: Harvard University Press, 1972), 5.

20. Douglas Peers, *Between Mars and Mammon: Colonial Armies and the Garrison State in India 1819–1835* (London: Tauris Academic Studies, 1995), 4–9.
21. For an analysis of military-fiscalism, or what Douglas Peers calls the 'Garrison State', see C. A. Bayly, 'The British Military-Fiscal State and Indigenous Resistance: India 1750–1820', *An Imperial State at War: Britain from 1689 to 1815*, ed. Lawrence Stone (London: Routledge, 1993); Douglas Peers, 'Gunpowder Empires and the Garrison State: Modernity, Hybridity, and the Political Economy of Colonial India, circa 1750–1860', *Comparative Studies of South Asia, Africa and the Middle East* 27, 2 (2007).
22. See Peers, *Between Mars and Mammon*, 125.
23. Ibid., 130.
24. The ratio of European to sepoy soldiers in the early nineteenth century varied from roughly 1:4 (in 1813) to 1:5 (around 1826). See ibid., 257.
25. *Report of the Commissioners Appointed to Inquire into the Sanitary State of the Army in India, with Abstract of Evidence* (London: H.M. Stationery Office, 1864), 21. For the men's pay, see C. J. Hawes, *Poor Relations: The Making of a Eurasian Community in British India, 1773–1833* (Richmond: Curzon, 1996), 12.
26. Patrick John Thomas, *The Growth of Federal Finance in India: Being a Survey of India's Public Finances from 1833 to 1939* (Madras: Oxford University Press, 1939), 50.
27. Seema Alavi, *The Sepoys and the Company: Tradition and Transition in Northern India, 1770–1830* (Delhi: Oxford University Press, 1995), 43.
28. Ideas of masculinity were central to this ideology. For a full examination of such constructions, see Mrinalini Sinha, *Colonial Masculinity: The 'Manly Englishman' and the 'Effeminate Bengali' in the Late Nineteenth Century* (Manchester: Manchester University Press, 1995).
29. Peers, 'Imperial Vice', 36.
30. Peers, 'Soldiers, Surgeons and the Campaigns to Combat Sexually Transmitted Diseases'' 139.
31. Ibid., 159.
32. This Act operated to varying degrees across the presidencies. The India Contagious Diseases Act was finally repealed in 1888, but, predictably, an unofficial version of the system carried on in many stations.
33. David Arnold, 'Medical Priorities and Practice in 19th Century British India', *South Asia Research* 5, 2 (1985); Harrison, *Public Health in British India.*
34. David Arnold, *Colonizing the Body: State Medicine and Epidemic Disease in Nineteenth Century India* (Berkeley: University of California Press, 1993), 203.
35. Rajnarayan Chandavarkar, 'Plague Panic and Epidemic Politics in India, 1896–1914', *Epidemics and Ideas: Essays on the Historical Perception of Pestilence*, 203–40 eds. Terence Ranger and Paul Slack (Cambridge: Cambridge University Press, 1992); Arnold, *Colonizing the Body*, 205.
36. These methods were utilised to counter a suspected plague outbreak in Pali in 1836. See Mark Harrison, *Climates & Constitutions: Health, Race, Environment and British Imperialism in India 1600–1850* (Oxford: Oxford University Press, 1999), 194.
37. Of course, this was not the case where 'public health' was concerned. See Harrison, *Public Health in British India*, 6.
38. Asia, Pacific and Africa Collections, British Library (hereafter APAC) Letter to Colonel W. Casement, C. B., Secretary to Government Military Department

from W. Burke Esq, Inspector General of Hospitals, His Majesty's Force in India, 21 April 1832. Correspondence with the Bombay Government respecting the abolition of Lock Hospitals at Bombay. Board of Commissioners Collections 1832–1833. APAC F/4/1338/53031.

39. David Washbrook, 'Progress and Problems: South Asian Economic and Social History, 1720–1860', *Modern Asian Studies* 21, 1 (1988), 79.
40. In contrast, there is a rich literature which focuses on the changes wrought on some of the individual groups of women later classified as 'prostitutes' (such as temple dancers and courtesans). See V. T. Oldenburg, 'Lifestyle as Resistance: The Case of the Courtesans of Lucknow, India', *Feminist Studies* 16 (1990), 259–288; Kunal Parker, '"A Corporation of Superior Prostitutes" Anglo-Indian Legal Conceptions of Temple Dancing Girls, 1800–1914', *Modern Asian Studies* 32, 3 (1998); Amrit Srinivasan, 'Reform and Revival: The Devadasi and Her Dance', *Economic and Political Weekly* 20, 44 (1985).
41. The *nautch* (as it was generally called by British observers) was a dance performed by specially trained dancers at royal courts, elite events and social gatherings.
42. David Washbrook also explores the role played by the colonial state in 'traditionalising' Indian society through legal practices. David Washbrook, 'Law, State and Agrarian Society in Colonial India', *Modern Asian Studies* 15, 3 (1981), 649–721. In a similar manner, Nicholas Dirks describes the role played by colonialism in the production of Indian 'tradition' Nicholas B. Dirks, *Castes of Mind: Colonialism and the Making of Modern India* (Princeton: Princeton University Press, 2001).
43. This book is indebted to recent studies on sexual and domestic slavery as well those that explore changes to concepts of gender as well as to 'family', 'marriage' and 'home'. See Tanika Sarkar, *Hindu Wife, Hindu Nation: Community, Religion and Cultural Nationalism* (New Delhi: Permanent Black, 2001); Partha Chatterjee, *The Nation and Its Fragments: Colonial and Postcolonial Histories* (Princeton, NJ: Princeton University Press, 1993); Aisika Chakrabarti, 'Widowhood in Colonial Bengal, 1850–1930' (unpublished PhD Thesis, University of Calcutta, 2004); Mytheli Sreenivas, *Wives, Widows, and Concubines: The Conjugal Family Ideal in Colonial India* (Bloomington: Indiana University Press 2008); Durba Ghosh, *Sex and the Family in Colonial India: The Making of Empire* (Cambridge: Cambridge University Press, 2006); Indrani Chatterjee, *Gender, Slavery and Law in Colonial India* (Delhi: Oxford University Press, 2002).
44. The perception that venereal disease originated in diseased women lasted well into the twentieth century. In his *Health Problems of the Empire*, Andrew Balfour asserted that it was an 'urban' malady, coming from where '... people congregate, where the prostitute, professional or otherwise, is busy ...' Andrew Balfour, *Health Problems of the Empire: Past Present and Future* (London: Collins, 1924), 308.
45. Philippa Levine, 'Rereading the 1890s: Venereal Disease as "Constitutional Crisis" in Britain and British India', *Journal of Asian Studies* 55, 3 (1996).
46. See Guy Attewell, *Refiguring Unani Tibb: Plural Healing in Late Colonial India* (Hyderabad: Orient Blackswan, 2007); Projit Bihari Mukharji, *Nationalizing the Body: The Medical Market, Print and Daktari Medicine* (London: Anthem, 2011). Mukharji rightly points out the danger, when exploring colonial

medical practices through the examination of government files or English-language texts produced, of seeing 'western' medical practice in India as a system. Ibid., 16. While this study examines the representations of European practitioners regarding venereal disease and drink, it points to a different conclusion, that of the many fractures within the system, while still highlighting where practitioners held more 'uniform' views (as well as the non-scientific reasons behind this united front).

47. There is an active debate regarding 'metropolitan' versus 'colonial' knowledge. It is beyond the scope of this book to suggest a broader theory to definitively describe how scientific 'knowledge' was produced in colonial India and the relationship of this knowledge to that being produced in Europe. However, it suggests that in terms of venereal disease control, European practitioners in India were central in the shaping of venereal disease control measures in the nineteenth century.

1 The East India Company, the army and Indian society

1. SM Edwardes, *Crime in India: A Brief Review of the More Important Offences Included in the Annual Criminal Returns with Chapters on Prostitution & Miscellaneous Matters* (London: Oxford University Press, 1924), 71. For references to the prostitute 'caste', see p. 79.
2. Lieutenant-Colonel James Marshall, 'Prevention of Venereal Disease in the British Army' (paper presented at the Inter-Allied Conference on War Medicine, Royal Society of Medicine, 1944), 260.
3. Herman de Watteville, *The British Soldier. His Daily Life from Tudor to Modern Times* (London: J. M. Dent & Sons, 1954), 78.
4. James notes that in 1786, nearly all of the 389 men of the 4th Bombay European Battalion gave their previous occupation as labourer, although there was a handful of craftsmen and butchers. Lawrence James, *Raj: The Making and Unmaking of British India* (London: Little, Brown and Company, 1997), 136. For conditions of service in the Company army, see Peter Stanley, *White Mutiny: British Military Culture in India, 1825–1875* (London: Hurst, 1998), 21.
5. See specifically Sections 292, 293, 372, 373 and 294, *Indian Penal Code*, 1860. These sections respectively deal with the buying and selling of minors for the purpose of prostitution, the sale of obscene books and objects and the commission of obscene acts and songs.
6. These were the so-called 'successor states' – rising regional powers who, while holding a titular allegiance to the Mughal emperor, were in effect, sovereign powers.
7. Bankey Bihari Misra, *The Central Administration of the East India Company, 1773–1834* (Manchester: Manchester University Press, 1959), 17. The Regulating Act itself was drafted as a parliamentary response to the Company's growing territorial and political power in India.
8. George Sir MacMunn and A. C. Lovett, *The Armies of India* (Bristol: Somerset Guides, 1911, 1984), 4.
9. For early reforms and a discussion of officer discontent, see Callahan, *The East India Company and Army Reform*, 121.

10. The three Presidency armies remained until 1895, when they were finally merged to form the Indian Army. For an engaging analysis of the mutiny among European soldiers following the absorption of the Company armies into the British Army, see Stanley, *White Mutiny*.
11. See Henry Yule, Arthur Burnell and W. Crooke, *Hobson-Jobson: A Glossary of Anglo-Indian Words and Phrases: And of Kindred Terms, Etymological, Historical, Geographical, and Discursive* (London: John Murray, 1886), 329; Stanley, *White Mutiny*, 7.
12. 'Sepoy' comes from the Persian 'sipahi', meaning 'soldier' – this was the term used to describe the Indian soldiers trained and dressed in a European style.
13. MacMunn and Lovett, *The Armies of India*, 4.
14. For the history of military recruitment in these areas (now known as Bihar, Varanasi and Awadh, respectively), see Dirk H. A. Kolff, *Naukar, Rajput, and Sepoy: The Ethnohistory of the Military Labour Market in Hindustan, 1450–1850* (Cambridge: Cambridge University Press, 1990), 180. For a detailed history of the Company's sepoys, see Alavi, *The Sepoys and the Company*; Omissi, *The Sepoy and the Raj*.
15. Edward M. Spiers, *The Army and Society 1815–1914* (London: Longman, 1980), 121.
16. Letter to the Honourable Warren Hastings, Governor General and Members of the Supreme Council from General Stubbert, Fort William, no. 12–14. 11 March 1782. NAI Foreign (Secret).
17. Omissi, *The Sepoy and the Raj*, 3.
18. Hawes, *Poor Relations*, 9.
19. Strachan, *Wellington's Legacy*, 52. For representations of the Irish soldier in India, see Alexander Bubb, 'The Life of the Irish Soldier in India: Representations and Self-Representations, 1857–1922', *Modern Asian Studies* 46, 4 (2012).
20. Hawes, *Poor Relations*, 12.
21. Ibid.
22. Strachan, *Wellington's Legacy*, 55.
23. Ibid., 64.
24. John Pearman and George Charles Henry Victor Paget Marquess of Anglesey, *Sergeant Pearman's Memoirs: Being, Chiefly, His Account of Service with the Third (King's Own) Light Dragoons in India, from 1845 to 1853, Including the First and Second Sikh Wars* (London: Jonathan Cape, 1968), 25.
25. PJ Marshall, 'British Society in India under the East India Company', *Modern Asian Studies* 31, 1 (1997), 93.
26. For an analysis of the development of 'martial races' theory, see Heather Streets-Salter, *Martial Races: the Military, Race and Masculinity in British Imperial Culture, 1857–1914* (Manchester: Manchester University Press, 2004).
27. David Cannadine, *Ornamentalism: How the British Saw Their Empire* (London: Allen Lane, 2001), 9.
28. Stanley, *White Mutiny*, 21.
29. Marshall, 'British Society in India under the East India Company', 93.
30. Ibid.
31. Hawes, *Poor Relations*, 12.
32. Sir Charles E. Trevelyan, K. C. B., *The British Army in 1868* (London: Longmans, Green and Co, 1868), 16.

33. Strachan, *Wellington's Legacy*, 53.
34. The original etymology of the term '*lal bazaar*' is unclear. One suggestion is that the phrase came from the '*lal kurti*' or red jackets which were the traditional uniforms of the British soldier. The second suggests that it was the standard association with a 'red light district' which gave them their name. Kenneth Ballhatchet suggests that while the term '*lal bazaar*' was not used formally until the nineteenth century, the process itself was already underway. I would argue that the term was used officially much earlier, however, as evidenced by Samuel Hickson's 1781 letter to his brother which describes the 'Loll Bazar' at Surat. See Ballhatchet, *Race, Sex and Class under the Raj*, 12 and letter from Samuel Hickson, 24 December 1781. Letters and other Papers Relating to Samuel Hickson of Market St Herts, Some Years of the East India Company Service and who Died 24 October 1814, Collected and Copied in the Year 1820 by W. E. Hickson 1781. APAC MSS EUR/B296/1.
35. Rev. James Long, *Selections from Unpublished Records of Government for the Years 1748 to 1767, Inclusive, Relating Mainly to the Social Condition of Bengal; with a Map of Calcutta in 1784* (Calcutta: Office of Superintendent of Government Printing, 1869), 68.
36. Suresh Chandra Ghosh, *The Social Condition of the British Community in Bengal 1757–1800* (Leiden: E. J. Brill, 1970), 76. It was only in the late eighteenth and early nineteenth century that the process of Catholic emancipation began in Britain.
37. Medical Board Proceedings, 30 April 1810, Re-establishment of Lock Hospitals at Bangalore, Wallajabad and Poonamallee Authorised. Board of Commissioners Collections 1813–1814. APAC F/4/379/9435.
38. Ibid.
39. Medical Board Proceedings, 1 May 1809, extracted in Resolutions of the Governor in Council. Madras Military Consultations, 11 May 1810. APAC P/256/67.
40. Ibid.
41. Betty Joseph, *Reading the East India Company, 1720–1840: Colonial Currencies of Gender* (Chicago: University of Chicago Press, 2004), 8.
42. Letter from Samuel Hickson, 24 December 1781. APAC MSS EUR/B296/1.
43. Ibid.
44. Ibid., 81. Although it is difficult to find an exact number for Hickson's battalion strength, the average sepoy battalion would have most likely been 300–500 men. Hickson, presumably is speaking solely of the Europeans in his group, which might have represented about 5–6 men. For approximate battalion strength, see Alavi, *The Sepoys and the Company*, 35. For European to Indian ratio, see Kaushik Roy, 'The Armed Expansion of the English East India Company, 1740s–1849', *A Military History of India and South Asia: From the East India Company to the Nuclear Era*, 1–15 eds. Daniel Marston and Chandar Sundaram (London: Praeger, 2007), 5.
45. Letter from Samuel Hickson, 24 December 1781. APAC MSS EUR/B296/1. Hickson himself was rather grumpy about the music and dancing, feeling himself unqualified 'either by abilities or inclination' to participate in either.
46. Pearman and Marquess of Anglesey, *Sergeant Pearman's Memoirs*, 108.
47. This was especially true when the mother herself was a slave. Indrani Chatterjee, 'Colouring Subalternity: Slaves, Concubines and Social Orphans',

Subaltern Studies X: Writings on South Asian History and Society, 49–97 eds. Gautam Bhadra, Gyan Prakash and Susie J. Tharu (New Delhi: Oxford University Press, 1999), 73.

48. Ghosh, *Sex and the Family in Colonial India*, 229.
49. David Arnold, 'European Orphans and Vagrants in India in the Nineteenth Century', *Journal of Imperial and Commonwealth History* 7, 2 (1979), 109–10.
50. Durba Ghosh, 'Making and Un-making Loyal Subjects: Pensioning Widows and Educating Orphans in Early Colonial India', *Journal of Imperial and Commonwealth History* 31, 1 (2003), 19.
51. Letter to Commander-in-Chief from Military Secretary, 31 December 1818. Bengal Military Collections 1818. APAC L/MIL/5/376, col. 3.
52. Ibid.
53. Ibid.
54. Pearman and Marquess of Anglesey, *Sergeant Pearman's Memoirs*, 66.
55. For an extensive study of these wills, as well as those of the Indian companions of European men, see Ghosh, *Sex and the Family in Colonial India*, chapters 3 & 4.
56. Will of Matthew Leslie, Probate Petitioned for on 20 February 1804, no. 13. Bengal Wills 1804. APAC L/AG/34/29/16.
57. Ibid.
58. Chatterjee, 'Colouring Subalternity'.
59. See 'British Siblings in Bihar in Search of Their Roots', *The Bihar Times*, 12 January 2009.
60. Baharampur.
61. Will of Thomas Naylor, Probate Granted 6 August 1782, no. 24. Bengal Wills 1780–1783. APAC L/AG/34/29/4.
62. Williamson himself was a fascinating figure. He held a wide range of interests, from music to soldiering, and operated in true entrepreneurial style. He departed India abruptly, having been suspended after (anonymously) publishing his opinion on Cornwallis' military policy in the *Calcutta Telegraph* in 1798. Despite publishing a number of books on his return to Europe, he died intestate in Paris in 1817. See Owain Edwards, 'Captain Thomas Williamson of India', *Modern Asian Studies* 14, 4 (1980), 675.
63. Captain Thomas Williamson, *The East India Vade-Mecum; or, Complete Guide to Gentlemen Intended for the Civil, Military, or Naval Service of the Hon. East India Company* (London: Black, Parry and Kingsbury, 1810), 414. Italics in original.
64. Ghosh, *The Social Condition of the British Community in Bengal*, 75.
65. Williamson, *The East India Vade-Mecum*, 413.
66. Ibid., 451.
67. Captain Bellew, *Memoirs of a Griffin; or, A Cadet's First Year in India* (London: William H. Allen and Co., 1843), 164.
68. Fanny Parkes, *Wanderings of a Pilgrim, in Search of the Picturesque, During Four-and-Twenty Years in the East; with Revelations of Life in the Zenana* (London: Pelham Richardson, 1850), vol. 1, 184.
69. Williamson, *The East India Vade-Mecum*, 456.
70. Ann Laura Stoler, 'Rethinking Colonial Categories: European Communities and the Boundaries of Rule', *Comparative Studies in Society and History* 31, 1 (1989), 147.
71. Hawes, *Poor Relations*, 64–5.

72. Ghosh, *The Social Condition of the British Community in Bengal*, 80.
73. Letter to Edward Wood, Secretary to Government from Robert Orme, Solicitor, Enclosing the Opinion of the Advocate General, 26 May 1817. Madras Military. Board of Commissioners Collections 1818–1819. APAC F/4/557/13666.
74. Maharashtra State Archives (hereafter MSA) Letter to the Secretary to Government, Military Department Bombay from the Adjutant General of the Army. 1 January 1868. MSA Military Department (1866–1868), vol. 1114.
75. Collection Regarding the Provision Made for Women and Orphan Children Left Destitute on the Departure of the King's Regiments. Military Secretary's Office, Bengal Military 1820. APAC L/MIL/5/376, col. 7.
76. Letter from Colonel Henry Torrens, 21 August 1820. Bengal Military Collections 1820. APAC L/MIL/5/376, col. 7.
77. Collection Regarding the Provision Made for Women and Orphan Children Left Destitute on the Departure of the King's Regiments. APAC L/MIL/5/376, col. 7.
78. Pearman and Marquess of Anglesey, *Sergeant Pearman's Memoirs*, 110.
79. Present-day Sabhra.
80. Pearman and Marquess of Anglesey, *Sergeant Pearman's Memoirs*, 60.
81. Letter to the Honourable Major General Sir Thomas Munro, Governor in Council, Draft General Order. Question as to the Allowances of the Wives Widows and Children of European Soldiers. Madras Military Collections 1825. APAC F/4/787/21363.
82. Minute from Thomas Munro, Governor in Council, November 1824. Madras Military Collections 1825. APAC F/4/787/21363.
83. Ibid.
84. Ghosh, 'Making and Un-Making Loyal Subjects', 3.
85. Ibid., 6.
86. Military Letter to Fort St George, 7 September 1808. Madras Military. Board of Commissioners Collections 1812. APAC F/4/360/8774.
87. Ibid.
88. Extract Military Letter to Madras, 24 August 1825. Country Born Widows of Soldiers Now Admitted to Clive's Fund. Half Castes to Be Called Indo-Britons. Madras Military. Board of Commissioners Collections 1829–1830. APAC F/4/1115/29907.
89. Draft Military Letter Madras, 29 July 1829. Madras Despatches, 6 May to 30 September 1829. APAC E/4/937.
90. Military Letter to Bengal, 24 May 1826. Bengal Despatches, 3 May to 27 September 1826. APAC E/4/717.
91. For this, see the collection of petitions from native and 'half caste' widows in Country-born Widows of Soldiers now Admitted to Clive's Fund. Military Secretary's Office. Madras Military Collections 1829. APAC F/4/1115/29907.
92. Ghosh, 'Making and Un-Making Loyal Subjects', 5.
93. Letter to the Honourable Mountstuart Elphinstone, President and Governor in Council from RH Hough, Military Auditor General's Office. 26 August 1826. MSA Military Department (1826), vol. 17.
94. In 1813, the Company's charter was reviewed and renewed. As part of this Regulating Act, the Company was required to open up its territories to missionaries.

95. Letter from the Secretary of the Bombay Bible Society, 8 November 1814. Bombay Public Consultations. Board of Commissioners Collections 1816–1817. APAC F/4/503/12032.
96. Ibid.
97. Ibid.
98. During the negotiations for the renewal of the Company's charter, one of the preconditions imposed on the Company was that they permit missionaries into its territory.
99. Council for World Mission Archives (hereafter CWM) Letter to Reverend Bogue from M. Hill. CWM North India, Bengal. Incoming Correspondence, Box 2, 1824–1829, Folder 1. 11 December 1824.
100. Ibid.
101. Officers and civil servants of the Company, on the other hand, had the resources to quietly continue their relationships for a longer period. It would appear that it was not until the early 1820s that such men started turning their backs on these relationships in larger numbers and many continued into the late 1850s.
102. See Dr William Acton, *Prostitution: Considered in Its Moral, Social & Sanitary Aspects in London and Other Large Cities with Proposals for the Control and Prevention of Its Attendant Evils* (London: John Churchill & Sons, 1857) and Michael Ryan, *Prostitution in London, with a Comparative View of that in Paris and New York* (London: H. Bailliere, 1839).
103. Nathaniel Halhed, ed. *A Code of Gentoo Laws, or, Ordinations of the Pundits, from a Persian Translation, Made from the Original, Written in the Shanscrit Language* (London: n.p., 1776), 171.
104. Banerjee, *Dangerous Outcast*, 177.
105. For the re-writing of 'tradition' and law in the early nineteenth century see Washbrook, 'Law, State and Agrarian Society in Colonial India'. For a similar transformation later in the century, see Dirks, *Castes of Mind*, 152.
106. Rimli Bhattacharya, 'The nautee in "the second city of the Empire"', *Indian Economic and Social History Review* 40, 2 (2003), 204. Katherine Schofield makes a similar argument about what she deems the 'liminality' of Mughal court musicians. See Katherine Schofield (nee Butler Brown), 'The Social Liminality of Musicians: Case Studies from Mughal India and Beyond', *Twentieth Century Music* 3, 1 (2007).
107. Chakrabarti, 'Widowhood in Colonial Bengal 32.
108. For an examination of the courtesan tradition in Lucknow, see Oldenburg, 'Lifestyle as Resistance and V. T. Oldenburg, *The Making of Colonial Lucknow, 1856–1877* (Princeton: Princeton University Press, 1984).
109. Oldenburg, *The Making of Colonial Lucknow*, 134.
110. C. A. Bayly, *Rulers, Townsmen and Bazaars: North Indian Society in the Age of British Expansion 1770–1870* (New Delhi: Oxford University Press, 2002), 269.
111. Ibid., 280.
112. Kokila Dang, 'Prostitutes, Patrons and the State', *Social Scientist* 21 (1993), 175.
113. Charu Gupta, *Sexuality, Obscenity, Community: Women, Muslims, and the Hindu Public in Colonial India* (New Delhi: Permanent Black, 2001), 111.
114. The word *devadasi* translates to 'handmaiden of the god'. For a history of the changing status of what she calls 'sacred prostitution', see Priyadarshini

Vijaisri, *Recasting the Devadasi: Patterns of Sacred Prostitution in Colonial South India* (New Delhi: Kanishka Publishers, 2004).

115. Awadh Kishore Prasad, *Devadasi System in Ancient India: A Study of Temple Dancing Girls of South India* (Delhi: H.K. Publishers and Distributors, 1990), 95.
116. Vijaisri, *Recasting the Devadasi*, 95.
117. Srinivasan, 'Reform and Revival: The Devadasi and Her Dance', 1869.
118. Vijaisri, *Recasting the Devadasi*, 45.
119. Janaki Nair, 'The Devadasi, Dharma and the State', *Economic and Political Weekly* 29, 50 (1994), 3158. Using the case of the *devadasi* in Mysore, Nair observes that the colonial state, in its attempts to control land revenue settlements, interfered with both the economic and cultural resources of the temple.
120. Pushpa Sundar, *Patrons and Philistines: Arts and the State in British India, 1773–1947* (Delhi: Oxford University Press, 1995), 243.
121. Francis Buchanan, *A Journey from Madras Through the Countries of Mysore, Canara, and Malabar* (London: Black, Parry, and Kingsbury, 1807), vol. II, 266. My italics.
122. An intermediary or moneylender. Kantababu was Hastings primary representative and a very successful and savvy businessman.
123. Somendra Candra Nandy, *Life and Times of Cantoo Baboo (Krisna Kanta Nandy): The Banian of Warren Hastings, 1742–1804* (Bombay: Allied Publishers, 1978), 447.
124. John H. Farrant, 'Grose, John Henry (b. 1732, d. in or after 1774)' *Oxford Dictionary of National Biography* (London: Oxford University Press, 2004).
125. John Henry Grose, *A Voyage to the East-Indies; Began in 1750: With Observations Continued Till 1764; Including Authentic Accounts of the Mogul Government in General ... of the European Settlements, etc.* (London: n.p., 1766), 138.
126. Ibid.
127. J. A. Dubois, *Description of the Character, Manners and Customs of the People of India: And of Their Institutions, Religious and Civil* (London: Longman, Hurst, Rees, Orme and Brown, 1817), 584.
128. Ibid., 586.
129. Eliza Fay, *Original Letters from India; Containing a Narrative of a Journey through Egypt, and the Author's Imprisonment at Calicut by Hyder Ally* (Calcutta: n.p., 1817), 229.
130. Ibid.
131. Sara Suleri, *The Rhetoric of British India* (Chicago, IL: University of Chicago Press, 1992), 6.
132. Emma Roberts, *Scenes and Characteristics of Hindostan, with Sketches of Anglo-Indian Society* (London: William H. Allen and Co., 1835), 247.
133. Reginald Heber, *A Narrative of a Journey through the Upper Provinces of India, from Calcutta to Bombay, 1824–1825* (London: John Murray, 1828), vol. 1, 47.
134. Ibid.
135. Letter to the Superintending Surgeon, Mysore Division, 24 December 1805. Board of Commissioners Collections 1807–1808. APAC F/4/200/4502.
136. General Orders by the Commander in Chief, 11 October 1805. Fort St George Military Consultations. Board of Commissioners Collections 1807–1808. APAC F/4/200/4502.

137. 'Prostitution in Relation to the National Health', *The Westminster Review* 92 (1869). In these demands, the medical authorities were never wholly successful, but they persisted in requesting that *devadasis* be included in the regulatory boundaries nonetheless.
138. Medical Board Proceedings, 30 April 1810. APAC F/4/379/9435.
139. Sagar.
140. Letter to James Jameson, Secretary to the Medical Board from Mr Dickson, Superintending Surgeon, Saugor Field Force. Bengal Medical Board Proceedings, no. 78, 9 June 1821. NAI Home (Medical).
141. Parker, 'A Corporation of Superior Prostitutes', 560.
142. Radhika Singha, *A Despotism of Law: British Criminal Justice and Public Authority in North India, 1772–1837* (Oxford: Oxford University Press, 1998), 147. For Singha, the significance of this legislation rests in the fact that it aimed to control the bodies of wives and daughters, essentially as property of the male householder and viewed women as lacking their own will or agency.
143. Parker, 'A Corporation of Superior Prostitutes', 601.
144. Chakrabarti, 'Widowhood in Colonial Bengal', 87.
145. Parker, 'A Corporation of Superior Prostitutes', 562.
146. Letter to Reverend Ellis from Mr Lacroix, Calcutta. CWM North India, Bengal. Incoming Correspondence, Box 5. 1837–1838, Folder 1, Jacket A. 20 October 1837. Underline in original.
147. This campaign combined both *nautch* and temple dancing and culminated in the passage of the Devadasi Abolition Bill in 1929–1931.
148. *Opinions on the Nautch Question Collected and Published by the Punjab Purity Association* (Lahore: New Lyall Press, 1894), 3.
149. Partha Chatterjee, *Nationalist Thought and the Colonial World: A Derivative Discourse?* (London: Zed for the United Nations University, 1986); Sarkar, *Hindu Wife, Hindu Nation*; Nair, 'The Devadasi, Dharma and the State'.
150. Minute by the Governor General, 29 October 1860. Proceedings, nos. 1–3, 3 November 1860. NAI Home (Education).
151. Establishment of Lock Hospitals, 1807. Board of Commissioners Collections 1805–7. APAC F/4/200/4502.
152. Parker, 'A Corporation of Superior Prostitutes', 561.

2 Regulating the body: experiments with venereal disease control, 1797–1831

1. *Report of the Commissioners Appointed to Inquire into the Sanitary State of the Army in India, with Abstract of Evidence* 21. In 1797, there were approximately 13,125 British soldiers serving in India. By the eve of the rebellion in 1857, it was estimated that this number had risen to 45,000. See Roy, 'The Armed Expansion of the English East India Company', 3; 'Chamber's Encyclopaedia, a Dictionary of Universal Knowledge for the People, Illustrated, vol. V', 645 (Edinburgh: W & R Chambers, 1863), 645. The men's pay represented just under one-fifth of this amount (using the figure of one shilling a day, the men earned just under £20 per annum).
2. 'Fevers' – often malarial, were increasingly seen as unfortunate side-effects of the 'tropics'. Earlier theories of climactic adjustment were falling out

of favour by the early nineteenth century. See Harrison, *Climates and Constitutions*; Arnold, *Colonizing the Body*.

3. *Annual Report on the Military Lock Hospitals of the Madras Presidency for the Year 1871* (Madras: Government Press, 1871).
4. Present-day Tiruchirapalli.
5. Extract, letter from Assistant Surgeon Price of HM 12th Foot at Trichinopoly [n.d.], Fort St George Military Consultations, 1 March 1805. Board of Commissioners Collections 1807–1808. APAC F/4/200/4502.
6. *Annual Report on the Military Lock Hospitals of the Madras Presidency for the Year 1871*.
7. These were the states and rulers who were persuaded or forced into 'alliances' with the Company.
8. James Scott's study of peasants in Southeast Asia labels such forms of passive resistance the 'weapons of the weak'. See James C. Scott, *Weapons of the Weak: Everyday Forms of Peasant Resistance* (New Haven: Yale University Press, 1985).
9. Chandavarkar, 'Plague Panic and Epidemic Politics in India', 122.
10. Harrison, *Public Health in British India*, 40.
11. William Dalrymple tells the story of Bombay President William Methwold who, in 1630, inadvertently admitted that Company workers in Surat had almost completely given up using Western medicines, instead preferring to take the advice of local doctors. He quotes Methwold as writing, 'wee for our parts doe hold that in things indifferent it is safest for an Englishman to Indianise, and so conforming himself in some measure to the diett of the country, the ordinarie phisick of the country will bee the best cure when any sicknesse shall overtake him'. See William Dalrymple, *White Mughals: Love and Betrayal in Eighteenth-Century India* (London: Viking, 2003), 21.
12. Arnold, 'Medical Priorities and Practice in 19th Century British India', 176. Ayurveda is an ancient Hindu healing practice based on the principles of balance and energy flow. Unani or Muslim 'traditional' medicine, is Persian in origin and combines Arab and Greek theories of balance and humours to explain bodily conditions.
13. Harrison, *Climates and Constitutions*, 3.
14. See Biswamoy Pati and Mark Harrison, eds. *Health Medicine and Empire: Perspectives on Colonial India* (Hyderabad: Orient Longman Limited, 2001); Harrison, *Climates and Constitutions*; Arnold, *Colonizing the Body*.
15. In the 1790s, Abernethy published a series of papers on anatomical subjects. As a result, he was elected a fellow to the Royal Society. In addition, he served as professor of anatomy and surgery at the Royal College of Surgeons from 1814 to 1817. See L. S. Jacyna, 'Abernethy, John (1764–1831)', *Oxford Dictionary of National Biography* (Oxford: Oxford University Press, 2004).
16. John Abernethy, *Surgical Observations on Diseases Resembling Syphilis: And on Diseases of the Urethra* (London: n.p., 1810), 4.
17. Ryan, *Prostitution in London*, 255.
18. Ibid., 257.
19. Levine, *Prostitution, Race, and Politics*, 70.
20. Josephine Butler, *Personal Reminiscences of a Great Crusade* (London: Horace Marshall & Son, 1898), 1. Restif de la Bretonne was a French novelist and the author of a number of tracts on social reform.
21. Long, *Selections from Unpublished Records of Government for the Years 1748 to 1767*, 454.

22. Ibid.
23. Myna Trustram notes that regular genital examination of soldiers was general throughout the British Army until 1859. See Trustram, 'Distasteful and Derogatory?'. However, there is little corresponding evidence to suggest that the same was true of Company soldiers in India.
24. Extract, Letter from Assistant Surgeon Price of HM 12th Foot at Trichinopoly [n.d.], Fort St George Military Consultations, 1 March 1805. APAC F/4/200/4502.
25. Ibid.
26. Thomas Goddard was named Commander-in-Chief of the forces at Bombay in 1781. As such, he led a number of campaigns against the Marathas.
27. Letter from Samuel Hickson, 24 December 1781. APAC MSS EUR/B296/1. My italics.
28. Enclosure, Letter to James Jameson, Secretary to the Bengal Medical Board from Mr Keys, Superintending Surgeon Berhampore, 8 May 1818. Bengal Medical Board Proceedings, 8 May to 20 June 1818. NAI Home (Medical).
29. Respectively, Kanpur, Danapur and Fatehgarh. Military Board to Government, 11 December 1797. Bengal Military Consultations, 1 to 22 December 1797. APAC P/19/37.
30. Letter to Military Board from Adjutant General, 24 June 1799, Enclosing General Regulations for Hospitals for the Reception of Diseased Women. Bengal Military Consultations, 1 to 30 July 1799. APAC P/19/57.
31. Ibid.
32. Extract, Letter from Assistant Surgeon Price of HM 12th Foot at Trichinopoly [n.d.], Fort St George Military Consultations, 1 March 1805. APAC F/4/200/4502.
33. *Annual Report on the Military Lock Hospitals of the Madras Presidency for the Year 1871.*
34. James, *Raj*, 138.
35. William Geddes, *Clinical Illustrations of the Diseases of India: As Exhibited in the Medical History of a Body of European Soldiers for a Series of Years from Their Arrival in that Country* (London: Smith, Elder & Co, 1846), 17.
36. A King's Officer (anon.), *Remarks on the Exclusion of Officers of His Majesty's Service from the Staff of the Indian Army' and on the Present State of the European Soldier in India, Whether as Regards His Services, Health, or Moral Character; with a Few of the Most Eligible Means of Modifying the One and Improving the Other, Advocated and Considered* (London: T & G Underwood, 1825), 87.
37. General Orders by the Commander in Chief, 11 October 1805. APAC F/4/200/4502.
38. Ibid.
39. This feature was repeated by all systems of sexual regulation that followed, including the Contagious Diseases Acts.
40. Fort St George Military Consultations, 1 March 1805. Board of Commissioners Collections 1807–1808. APAC F/4/200/4502.
41. Medical Board Proceedings, 25 March 1805. Board of Commissioners Collections 1807–1807. APAC F/4/200/4502.
42. Ibid.
43. Letter to the Superintending Surgeon, Mysore Division, 24 December 1805. APAC F/4/200/4502.

44. Proceedings, nos. 112–115(A), 20 February 1869. NAI Home (Public).
45. Letter to John Murray MD, Deputy Inspector General of Hospitals, Madras from T. Lewis MD, Surgeon 4th King's Own, 14 June 1838. Madras Military Consultations, 21 August to 4 September 1838. APAC P/267/10. My italics.
46. Report from the Medical Board and Minute Thereon, 19 November 1813. Madras Military Consultations, 13 to 26 November 1813. APAC P/258/27.
47. Pratik Chakrabarti points out that by the middle of the eighteenth century, 'Country' medicines formed part of the standard supply for the medical establishment of Madras. See Pratik Chakrabarti, '"Neither of Meate nor Drinke, but What the Doctor Alloweth": Medicine Amidst War and Commerce in Eighteenth-Century Madras', *Bulletin of the History of Medicine* 80 (2006), 10.
48. Letter to the Military Board from the Military Secretary, 24 June 1799. Bengal Military Consultations, 1 to 30 July 1799. APAC P/19/57.
49. An Account of Country Medicine Expended from 15 to 31 December 1797 for Two Women. Bengal Military Consultations, 1 to 30 July 1799. APAC P/19/57. The term 'country' medicine was used to imply any of the myriad 'locally' sourced medicines – that is anything that did not come from Europe. However, as Pratik Chakrabarti has argued, this was an inexact term that could indicate medicines and materials with a geographically broad provenance. See Pratik Chakrabarti, *Materials and Medicine: Trade, Conquest and Therapeutics in the Eighteenth Century* (Manchester: Manchester University Press, 2010).
50. Measures for the Better Preservation of the Health of the European Soldiery, 1829. Bengal Military. Board of Commissioners Collections 1829–1830. APAC F/4/1079/29310.
51. Present-day Kheda.
52. A notable exception to this pattern can be seen in the table below.

Venereal disease admissions during field expedition in Bombay, April 1817

Corps	**Number of**		**Grand Total**
	Officers	**Privates**	
His Majesty's 17th Dragoons	2	6	8
His Majesty's 47th Regiment	3	3	6
H.C. Artillery and Lascars	–	1 Eur & 15 Lascars	16
Grenadier Battalion	1	8	9
Flank Battalion	1	37	38
1 Battalion 8th Regiment	..	15	15
2nd Battalion 8th Regiment	..	9	9
Cavalry & Pioneers	..	6	6
Commissariat	..	5	5
Total	**7**	**105**	112

Source: Bombay Military Consultations, 14 May 1817. APAC F/4/563/13819.

53. One exception to this seems to be the suggestion made in 1810 by William Ingledew, surgeon at Mysore, that '... out of one hundred [native] men, who have attained the middle age, there are not more than ten, who are perfectly free from it, and that all the rest, have their habits more or less tainted with that poison'. Ingledew does not, however, provide any further evidence to substantiate his claim. See Treatise on Venereal Disease, 6 February 1810. Madras Military. Board of Commissioners Collections 1812–1813. APAC F/4/345/8031.
54. Lock Hospitals: The Superintending Surgeon Reports. Fort St. George Military Consultations, 9 May 1809. Madras Military. Board of Commissioners Collections 1812–1813. APAC F/4/345/8032.
55. Fort St George Military Consultations, 1 March 1805. APAC F/4/200/4502.
56. Medical Board Proceedings, 16 November 1813. Madras Military. Board of Commissioners Collections 1815–1816. APAC F/4/486/11710.
57. Ibid.
58. Lock Hospitals: The Superintending Surgeon Reports. APAC F/4/345/8032.
59. Minutes of Commander-in-Chief, 24 September 1805. Board of Commissioners Collections 1807–1808. APAC F/4/200/4502.
60. The numbers for lock hospital admissions at Bangalore are especially striking. Without surviving records for that hospital, however, it is difficult to determine more specifically what the women were admitted for (and indeed the reason for the high level of deaths).
61. Extract Bombay Military Consultations, 2 September 1808. Establishment of a Lock Hospital at Kaira and Baroda. Board of Commissioners Collections 1818–1819. APAC F/4/563/13819.
62. Letter to James Jameson, Secretary to Medical Board from Mr Hamilton, Superintending Surgeon Chunar, 15 April 1820. Bengal Medical Board Proceedings, 20 May 1820. NAI Home (Medical).
63. Fort St George Military Letter, 6 February 1810. Madras Military. Board of Commissioners Collections 1812–1813. APAC F/4/345/8032. Note – one pagoda was equal to eight shillings. Therefore, this represented the not insignificant cost of £1,200.
64. Tamil Nadu State Archives (hereafter TNSA) Letter to the Senior Surgeon Present in the Provinces of Malabar and Canara from Mr Mackenzie, Acting Secretary Medical Board. 30 June 1809. TNSA Surgeon General's Records, vol. 17.
65. Letter to Board of Commissioners from Fort St George, 15 March 1811. Madras Military. Board of Commissioners Collections 1813–1814. APAC F/4/379/9435.
66. Ibid.
67. Ibid.
68. Bombay Military Consultations, 31 December 1812. Board of Commissioners Collections 1818–1819. APAC F/4/563/13819.
69. Ibid.
70. Medical Board Proceedings, 30 April 1810. APAC F/4/379/9435.
71. Extract, letter to the Medical Board from the Secretary to the Government, Military Department. 16 June 1809. TNSA Surgeon General's Records, vol. 17.
72. Medical Board Proceedings, 16 November 1813. APAC F/4/486/11710.

73. Letter to Mr Reddie from Mr Mansell, Garrison Surgeon Allahbad, 18 April 1821. Bengal Medical Board Proceedings, 26 April 1821. NAI Home (Medical).
74. Letter to Colonel Casement from James Jameson, Secretary to the Medical Board, 26 March 1819, no. 124, Medical Board Proceedings, 13 April 1819. NAI Home (Medical).
75. Ibid.
76. The official hospital at the station was not actually constructed until 1815.
77. Letter to John Adam, Secretary to the Medical Board from AB Webster, Assistant Surgeon, 2nd Extra Regiment, 6 March 1827. Bengal Medical Proceedings, 29 March 1827. NAI Home (Medical).
78. Present-day Almora.
79. Letter to John Adam, Secretary to the Medical Board from AB Webster, Assistant Surgeon, 2nd Extra Regiment, 6 March 1827. NAI Home (Medical).
80. Ibid.
81. Ibid.
82. Respectively, Palayamkottai, Solapur and Walajabad.
83. Madras Military Proceedings, 15 April 1829. Madras Despatches, 3 September 1828 to 15 April 1829. APAC E/4/936.
84. Letter to Lieutenant Simons, Establishment of Lock Hospitals. Board of Commissioners Collections 1807–1808. APAC F/4/200/4502.
85. Lock Hospitals: The Superintending Surgeon Reports. APAC F/4/345/8032. This argument closely mirrors that made by prison observers, where officials alleged that the poor referred to prisons as 'our father-in-law's house' – a place where, in times of hardship or famine, food and shelter would be provided. See David Arnold, 'The Colonial Prison: Power, Knowledge and Penology in Nineteenth-Century India', *Subaltern Studies VIII: Essays in Honour of Ranajit Guha*, 148–57 eds. David Arnold, David Hardiman and Ranajit Guha (Delhi: Oxford University Press, 1994), 168. For the ways in which women in Madras utilised the lock hospitals to their benefit during later periods of famine, see Hodges, '"Looting" the Lock Hospital'.
86. Extract Military Letter from Fort St. George, 4 February 1814. Madras Military. Board of Commissioners Collections 1815–1816. APAC F/4/486/11710.
87. The Military Board does not appear to have followed this advice, however, perhaps conscious of the need to cut costs. By 'pariah' women, the Board was suggesting those *dalit* women who were supposedly outside the caste system.
88. Letter to Mr Adam, Secretary to Bengal Medical Board from Mr Langstaff, Superintending Surgeon Meerutt, 22 April 1827. Bengal Medical Board Proceedings, 7 May 1827. NAI Home (Medical).
89. Letter to Colonel Casement from James Jameson, Secretary to the Medical Board, 26 March 1819. NAI Home (Medical).
90. For a discussion of resistance to vaccination, see Niels Brimnes, 'Variolation, Vaccination and Popular Resistance in Early Colonial South India', *Medical History* 48 (2004).
91. Lock Hospitals: The Superintending Surgeon Reports. APAC F/4/345/8032.
92. Ibid.
93. Proceedings of the Medical Board Recommending a Lock Hospital at the Presidency, St Thomas' Mount, 11 September 1815. Madras Military Consultations, 3 to 16 September 1815. APAC P/259/6. My italics.

94. A slightly more balanced report that followed the closure of the lock hospitals noted that the women sought treatment from hakims, who treated them with 'mercury and vegetable compounds'. See Letter to John Murray, Deputy Inspector General of Hospitals Her Majesty's Forces, Madras from Robert Davis, Assistant Surgeon HM 39th Regiment, Bangalore, 30 June 1838. Madras Military Consultations, 21 August to 4 September 1838. APAC P/267/10.
95. Letter to John Murray MD, Deputy Inspector General of Hospitals, Madras from T. Lewis MD, Surgeon 4th King's Own, 14 June 1838. APAC P/267/10. My italics.
96. Letter to Colonel Casement from James Jameson, Secretary to the Medical Board, 26 March 1819. NAI Home (Medical).
97. Letter to John Murray, Deputy Inspector General of Hospitals, from AB Morgan, Assistant Surgeon in Medical Charge, 57th Foot [n.d.]. Madras Military Consultations, 21 August 1838. APAC P/267/10.
98. Enclosure, letter to James Jameson, Secretary to the Bengal Medical Board from Mr Keys, Superintending Surgeon Berhampore, 8 May 1818. NAI Home (Medical).
99. Rajnarayan Chandavarkar, 'Customs of Governance: Colonialism and Democracy in Twentieth Century India', *Modern Asian Studies* 41, 3 (2007), 449.
100. Singha, *A Despotism of Law*, 11.
101. Radhika Singha has argued that 'order' and 'justice' in the Mughal period largely rested on the discretionary powers vested in *nazims, faujdars* and *kotwals* (respectively the governors, military commanders and police chiefs who worked together in each district). See ibid., 1–35.
102. Report from the Medical Board and Minute, 19 November 1813. APAC P/258/27.
103. Letter to the Right Honourable H. Elliot, Governor in Council from the Military Board, 11 September 1815. Madras Military Consultations, 3 to 16 September 1815. APAC P/259/6.
104. Extract, Letter to Captain Sterling, Adjutant General, 24 October 1831. Governor General's Minute. Bengal Military Consultations, 6 to 30 January 1832. APAC P/34/16.
105. *Batta* was the allowance given to officers and soldiers serving in the field.
106. Letter to Colonel Thomas, Commanding Belgaum from Lieutenant Colonel Green, Commanding HM 20th. Correspondence with the Bombay Government respecting the abolition of Lock Hospitals at Bombay. Board of Commissioners Collections 1832–1833. APAC F/4/1338/53031.
107. Present-day Pune.
108. Letter to the Acting Assistant Adjutant General, Poona Division from Colonel Sullivan, 29 June 1833. Bombay Military Consultations, 17 July 1833. APAC P/360/10.
109. Letter to James Jameson, Secretary Medical Board from Mr Gibb, Superintending Surgeon, 2nd Division Field Army Meerutt, 9 February 1819. Bengal Medical Board Proceedings, 17 March 1819. NAI Home (Medical).
110. The Bengal Military Board, reluctant to interfere judicially, refused a similar request in 1819 to extend the 'surveillance' methods in place, stating that such '... regulations, affecting women in the vicinity of the cantonments

who are under civil surveillance cannot be disturbed'. See Military letter to Bengal, 15 September 1824. Bengal Despatches, 4 August to 4 November 1824. APAC E/4/712.

111. Present-day Rajkot.
112. Letter to the Officer Commanding at Rajcote from P McDowell, Assistant Surgeon, 26 June 1825. Bombay Military Consultations, 7 September 1825. APAC P/358/34.
113. Letter to the Assistant Adjutant General from G Gordon, Major Commanding Rajcote, 27 June 1825. Bombay Military Consultations, 7 September 1825. APAC P/358/34.
114. Bombay Military Consultations, 18 October 1826. APAC P/358/50/9–12.
115. See, for example, Mr Reddie's returns of camp followers in the Medical Board index, 12 September 1823. Medical Board Proceedings. 1 October 1823. NAI Home (Medical).
116. Letter to Mr Adam, Secretary to Bengal Medical Board from Mr Langstaff, Superintending Surgeon Meerutt, 22 April 1827. NAI Home (Medical).
117. Enclosure, letter to James Jameson, Secretary to the Bengal Medical Board from Mr Keys, Superintending Surgeon Berhampore, 8 May 1818. NAI Home (Medical).
118. Ibid.
119. For such an example in Poona in 1833, see Letter to the Acting Assistant Adjutant General from Colonel Sullivan, 29 June 1833. APAC P/360/10.
120. Proceedings of the Medical Board, 11 September 1815. APAC P/259/6.
121. Letter to Colonel Casement from W Burke, 21 April 1832. APAC F/4/1338/53031.
122. Scott points out that most commonly, such resistance 'takes the form of passive non-compliance, subtle sabotage, evasion and deception'. Scott, *Weapons of the Weak*, 31.

3 Medicine and disease in the 'age of reform'

1. The term 'Anglicists' is used to describe the loosely knit alliance of liberal, evangelical utilitarian thinkers who pushed for changes to the way India was governed – to be 'Anglicised' – in among other things, the language of law and education. This group is portrayed in opposition to the 'Orientalists' – those who adopted a conservative approach to rule and believed that India should be governed by its own laws, 'traditions' and institutions. See Eric Stokes, *The English Utilitarians and India* (Delhi: Oxford University Press, 1989); Javed Majeed, *Ungoverned Imaginings: James Mill's The History of British India and Orientalism* (Oxford: Clarendon Press, 1992); John Rosselli, *Lord William Bentinck: The Making of a Liberal Imperialist, 1774–1839* (London: Chatto and Windus, 1974). A more recent notable exception to this comes with Ishita Pande's recent work that highlights multivalent 'Orientalisms' and explores the role of medicine and science in shaping liberalism. See Ishita Pande, *Medicine, Race and Liberalism in British Bengal: Symptoms of Empire* (London: Routledge, 2010).
2. See Lucy Carroll, 'Law, Custom and Statutory Social Reform: "The Hindu Widow Remarriage Act of 1856"', *Indian Economic and Social History Review* 20, 4 (1983); Lata Mani, *Contentious Traditions: The Debate on Sati in Colonial India* (Berkeley: University of California Press, 1998).

3. For an examination of another perceived bodily 'threat' to the Company – namely 'thugee', see Radhika Singha, '"Providential" Circumstances: The Thuggee Campaign of the 1830s and Legal Innovation', *Modern Asian Studies* 27, 1 (1993); Kim A. Wagner, *Thuggee: Banditry and the British in Early Nineteenth-Century India* (Basingstoke: Palgrave Macmillan, 2007).
4. For a study of the ideology and hierarchy of the Indian Civil Service and its Civilians, see Clive Dewey, *Anglo-Indian Attitudes: The Mind of the Indian Civil Service* (London: Hambledon, 1993). For further detail, see Kathryn Tidrick's discussion of Henry and John Lawrence and the so-called 'Punjab school' in Kathryn Tidrick, *Empire and the English Character* (London: Tauris, 1990).
5. In a similar manner, Pratik Chakrabarti identifies three phases in the trajectory of early British medicine in India in the eighteenth century. However, Chakrabarti's analysis focuses almost wholly on the period before these services were 'professionalised'. Chakrabarti, *Materials and Medicine: Trade, Conquest and Therapeutics in the Eighteenth Century*, 83.
6. For an examination of these Indian medical colleges and hospitals, see Mark Harrison, 'Was There an Oriental Renaissance in Medicine? The Evidence of the Nineteenth-Century Medical Press', *Negotiating India in the Nineteenth-Century Media*, 233–53 eds. David Finkelstein and Douglas M. Peers (Basingstoke: Macmillan, 2000), 237–8.
7. For a discussion on the earlier transformation in the field of medicine, health and medical practice in the eighteenth and early nineteenth centuries, see Christopher Lawrence, *Medicine in the Making of Modern Britain, 1700–1920* (London: Routledge, 1994), esp. chapter 2.
8. David C. Potter, *Government in Rural India. An Introduction to Contemporary District Administration* (London: London School of Economics & Political Science; G. Bell & Sons, 1964), 5.
9. J. M. Bourne, 'The East India Company's Military Seminary, Addiscombe, 1809–1858', *Journal of the Society for Army Historical Research* 57 (1979), 207.
10. Ibid., 214.
11. Anil Seal analyses the implications of the continued move toward more formal bureaucratic and administrative centralisation and training in the second half of the nineteenth century. See Anil Seal, *The Emergence of Indian Nationalism: Competition and Collaboration in the Later Nineteenth Century* (Cambridge: Cambridge University Press, 1971), esp. 1–24.
12. Report of the Committee appointed to Enquire into the Plan for Forming an Establishment at Home for the Education of Young Men Intended for the Company's Civil Service in India. 26 October 1804. APAC J/2/1.
13. The entrance examination for the college progressively grew more difficult, to the point at which, during the 1830s, a prospective student was required to prove his knowledge not only of the Gospels, but be able to translate portions of Homer, Herodotus and Sophocles, as well as a number of Latin authors including Cicero, Tacitus and Virgil. See the 1838 *East India Register and Directory*, cited in H. M. Stephens, 'The East India College at Haileybury', *Colonial Civil Service. The Selection and Training of Colonial Officials in England, Holland, and France ... With an Account of the East India College at Haileybury, 1806–1857*, 233–346 ed. Abbott Lawrence Lowell (New York: Macmillan, 1900), 287.
14. Rosane Rocher, 'Sanskrit for Civil Servants 1806–1818', *Journal of the American Oriental Society* 122, 2 (2002), 387.

15. John Beames, *Memoirs of a Bengal Civilian: The Lively Narrative of a Victorian District Officer* (London: Eland, 1984), 63–4.
16. Report of the Committee on an Establishment ... for the Civil Service in India. APAC J/2/1.
17. Beames, *Memoirs of a Bengal Civilian*, 63.
18. 'Rules and Regulations of the Honourable East India Company's Seminary at Addiscombe, 1834', *Calcutta Review* 2 (1844), 137.
19. Bourne, 'The East India Company's Military Seminary', 208.
20. 'Rules and Regulations of the Honourable East India Company's Seminary at Addiscombe, 1834', 130.
21. Bourne, 'The East India Company's Military Seminary', 208.
22. 'Rules and Regulations of the Honourable East India Company's Seminary at Addiscombe, 1834', 141.
23. On the production of a letter of invitation from a near-by friend or relative, Addiscombe's cadets were granted permission to cadets to leave the Seminary from Saturday afternoon to Monday morning. See ibid., 130.
24. Beames, *Memoirs of a Bengal Civilian*, 68.
25. D. G. Crawford, *A History of the Indian Medical Service, 1600–1913* (London: W. Thacker & Co., 1914), 509.
26. Ibid., 197.
27. 'Editorial', *The India Journal of Medical and Physical Science* 1, 1 (1836), 17.
28. Crawford, *A History of the Indian Medical Service, 1600–1913*, 214. This number failed to account for the legions of Indian *vaids*, *hakims*, apothecaries and assistants, who ministered to the needs of both European and sepoy troops.
29. Lieutenant-Colonel Donald Macdonald, *Surgeons Twoe and a Barber. Being Some Account of the Life and Work of the Indian Medical Service, 1600–1947* (London: William Heinemann Medical Books, 1950), 109.
30. It took a further 36 years until the London School of Tropical Medicine opened at the Albert Dock Seamen's Hospital.
31. Crawford, *A History of the Indian Medical Service, 1600–1913*, 370.
32. Ibid., 367.
33. C. A. Bayly, *Indian Society and the Making of the British Empire: The New Cambridge History of India* (Cambridge: Cambridge University Press, 1988), 98.
34. Peers, *Between Mars and Mammon*, 123.
35. Ibid., 132.
36. Bayly, *Indian Society and the Making of the British Empire*, 121.
37. Peers, *Between Mars and Mammon*, 189–92.
38. Agency houses served a number of financial intermediary roles, for example, they acted as bankers for the European community in India and agents investing on behalf of European companies. By the early nineteenth century, a small number of agency houses had monopolised large sections of the Indian economy, disrupting and displacing indigenous trade networks. For more on the financial collapse of the 1820s, see Amales Tripathi, *Trade and Finance in the Bengal Presidency, 1793–1833* (Calcutta: Oxford University Press, 1979), 170.
39. C. H. Philips, ed. *The Correspondence of Lord William Cavendish Bentinck, Governor-General of India 1828–1835* (Oxford: Oxford University Press, 1977), xvi.

40. Tripathi, *Trade and Finance in the Bengal Presidency*, 178.
41. Peers, 'Soldiers, Surgeons and the Campaigns to Combat Sexually Transmitted Diseases', 152.
42. Separate Military Letter from Bengal, 29 May 1832. Correspondence with the Bombay Government Respecting the Abolition of Lock Hospitals at Bombay. Board of Commissioners Collections 1832–1833. APAC F/4/1338/53031.
43. Harrison, 'Was There an Oriental Renaissance in Medicine?', 237.
44. Pande, *Medicine, Race and Liberalism in British Bengal*, 68.
45. Arnold, *Science, Technology and Medicine*, 63.
46. Harrison, 'Was There an Oriental Renaissance in Medicine?', 238.
47. Arnold, 'Medical Priorities and Practice in 19th Century British India', 176.
48. Frederick Corbyn, 'Medical College – Defects in its Constitution', *The India Journal of Medical and Physical Science* 1, 4 (1836), 187.
49. While late eighteenth and early nineteenth century descriptions of Indian practitioners refer to the men as 'doctors', by the 1830s, this term was replaced with less senior titles. This is not to suggest that these practitioners were any less experienced than their European counterparts. For more recent studies of South Asian practitioners of 'western' medicine see Projit Bihari Mukharji, *Nationalizing the Body: The Medical Market, Print and Daktari Medicine* (London: Anthem, 2009); Mridula Ramanna, *Western Medicine and Public Health in Colonial Bombay* (London: Sangam, 2002).
50. Mukharji, *Nationalizing the Body: The Medical Market, Print and Daktari Medicine*, 5.
51. Poonam Bala, *Medicine and Medical Policies in India: Social and Historical Perspectives* (Lanham: Lexington Books, 2007), 74.
52. In one example, the Dispensary reports for 1848 proudly note that 'Ruheemood Deen, Mesmeriser' was elected to the medical staff of the Native General Hospital in that year. See Half Yearly Reports of the Government Charitable Dispensaries Established in the Bengal and North Western Provinces ending 31 March 1848, 104. APAC V/24/736.
53. Pamela Gilbert, *Mapping the Victorian Social Body* (Albany: State University of New York Press, 2004), 160.
54. For detailed examination of the development of Victorian racial theories, see Michael Banton, *Racial Theories* (Cambridge: Cambridge University Press, 1998); George W. Stocking, *Victorian Anthropology* (New York: Free Press, 1987); Nancy Stepan, *The Idea of Race in Science: Great Britain 1800–1960* (London: Macmillan, 1982).
55. Blumenbach is often credited with developing the science of comparative anatomy. *On the Natural Variety of Man* was soon translated into German, French and English. Blumenbach found a hugely supportive audience in Britain's scientific community. In the third edition of his *Treatises*, he directly addresses Sir Joseph Banks, the then President of the Royal Society in London.
56. Johann Friedrich Blumenbach, 'On the Natural Varieties of Mankind', *The Anthropological Treatises of Johann Friedrich Blumenbach*, ed. Thomas Bendyshe (Boston: Milford House, 1865), 101.
57. Ibid., 156.
58. Stocking, *Victorian Anthropology*, 48.
59. Stepan, *The Idea of Race in Science: Great Britain 1800–1960*, 36. See Charles Darwin, *On the Origin of Species by Means of Natural Selection; or, The*

Preservation of Favoured Races in the Struggle for Life (London: CRW Publishing, 1859, 2004), 388–89.

60. Arnold, 'Medical Priorities and Practice in 19th Century British India', 168.
61. James Johnson, following his years as an apprentice and apothecary, began his career as a surgeon's mate (and later surgeon) in the navy. As such, he accompanied expeditions to Newfoundland, Gibraltar, Egypt and India (among other points). His first work on India, *The Oriental Voyager*, was published in 1807. He would go on to serve as both editor of the influential *Medico-Chirurgical Review* and physician to the William IV. For his early India observations, see James Johnson, *The Oriental Voyager; or, Descriptive Sketches and Cursory Remarks, on a Voyage to India and China ... in the years 1803–4–5–6* (London: James Asperne, 1807).
62. Mark Harrison, '"The Tender Frame of Man": Disease, Climate, and Racial Difference in India and the West Indies, 1760–1860', *Bulletin of the History of Medicine* 70, 1 (1996), 79.
63. Benjamin Mosely, quoted in James Johnson, *The Influence of Tropical Climates, More Especially the Climate of India, on European Constitutions* (London: Stockdale, 1813), 477. Emphasis in original.
64. Ibid., 479.
65. Ibid., 478.
66. Robert John Thornton, *The Medical Guardian of Youth; or, A Popular Treatise on the Prevention and Cure of the Venereal Disease, so that Every Patient May Act for Himself: Containing the Most Approved Practice, and the Latest Discoveries in This Branch of Medicine* (London: n.p., 1816), 16.
67. Ibid.
68. Present-day Sirmur.
69. Letter to Mr Jameson, Secretary to the Bengal Medical Board from Mr Ledman, Assistant Surgeon Sermoor Battalion, 16 January 1819. Medical Board Proceedings, 12 February 1819. NAI Home (Medical).
70. Ibid. So convinced was Ledman of the moral decay prevalent in the hills surrounding Deyrah, that while he suggested that a lock hospital system would reduce venereal numbers if instituted, he noted the improbability of ever fully eradicating the disease.
71. In an interesting twist, John Helfer, writing of the people of Tenasserim, noted of the offspring of Burmese women and European men that, 'We have not yet any proof how children by English fathers and Burmese mothers will turn out when grown up, the intercourse between the two nations having subsisted but 14 years; if we however may judge from what the children promise at present, we should be inclined to anticipate that they will be superior to the progeny of Europeans by Indian women'. John William Helfer, 'Third Report on Tenasserim – the surrounding Nations, – Inhabitants, Natives and Foreigners, – Character, Morals and Religion', *Journal of the Asiatic Society* VIII (1839), 1000.
72. O. P. Kejariwal, *The Asiatic Society of Bengal and the Discovery of India's Past, 1784–1838* (Oxford: Oxford University Press, 1988), 39.
73. Kapil Raj, 'The Historical Anatomy of a Contact Zone: Calcutta in the Eighteenth Century', *Indian Economic and Social History Review* 48, 1 (2011), 73.
74. Benedict Anderson, *Imagined Communities: Reflections on the Origin and Spread of Nationalism* (London: Verso, 1983).

75. Mrinalini Sinha, 'Britishness, Clubbability, and the Colonial Public Sphere: The Genealogy of an Imperial Institution in Colonial India', *The Journal of British Studies* 40, 4 (2001), 491.
76. *Transactions of the Medical and Physical Society of Calcutta* (Calcutta: Messrs Thacker & Co, 1825), iv.
77. 'Editorial', *Gleanings in Science* 1 (1829), 13.
78. In addition to arguing for the truth of extinction, Cuvier worked on systems of classification for vertebrate and invertebrate life forms, dividing animal life into four 'branches'.
79. James Annesley, *Sketches of the Most Prevalent Diseases of India, Comprising a Treatise on Epidemic Cholera of the East* (London: n.p., 1825).
80. 'Editorial', 278 and 342. Tytler took charge of the Native Medical Institution in 1827, having already taught at both the Sanskrit and Mahommedan Colleges.
81. In Prichard's case, as Ishita Pande points out, he formulated some of his ideas on race and caste as a result of a chance encounter with the Rammohan Roy. See Pande, *Medicine, Race and Liberalism in British Bengal*, 33.
82. Andrew Thompson, *The Empire Strikes Back? The Impact of Imperialism on Britain from the Mid-Nineteenth Century* (Harlow: Longman, 2005).
83. David Arnold, 'Race, Place and Bodily Difference in Early Nineteenth-Century India', *Historical Research* 77, 196 (2004), 257.
84. For instance, the circular sent to Surgeons to examine the treatment of venereal disease in 1831 questioned 'With reference to the constitution, mode of life, and diseases which prevail among the natives in the different provinces of India, do you consider them more or less fitted for the mercurial mode of treatment than the natives of our own country?' Circular letter to the Superintending Surgeons of Divisions, 14 October 1831. Bengal Medical Board Proceedings, 17 October 1831. NAI Home (Medical).
85. George Cuvier, a French naturalist, was instrumental in developing the field of comparative anatomy. Pieter Camper, a Dutch physician and anatomist, developed the idea of the 'facial angle'. Franz Joseph Gall, developed 'cranioscopy' or phrenology, which asserted that mental and moral capabilities could be determined by studying the shape of a person's skull.
86. 'Medical Jurisprudence – Phrenology', *The India Journal of Medical and Physical Science* 1, 1 (1834), 18.
87. Ibid.
88. Cally Coomar Doss, *The Pamphleteer* 1, 1 (1850), 1.
89. 'Phrenological Development of the Bengalees', *The Pamphleteer* 1, 3 (1850), 66.
90. Ibid., qualifying this further, he insisted that such traits were especially identifiable in the inhabitants of eastern Bengal. Clearly, the author's division of Bengalis would appear to reflect the biases of his own background.
91. Ibid., 68.
92. 'Queries Respecting the Human Race, to Be Addressed to Travellers and Others. Drawn Up by a Committee of the British Association for the Advancement of Science, Appointed in 1839, and Circulated by the Ethnographical Society of London', *Journal of the Asiatic Society* 13 (1844), 919–32.
93. Kejariwal, *The Asiatic Society of Bengal*, 194.

94. Brian Houghton Hodgson, 'Tibetan Type of Mankind', *Journal of the Asiatic Society* XVII, 2 (1848), 222. A monograph which followed on the 'aborigines' of India (the preface of which was re-printed in the *Journal of the Asiatic* Society in 1849), assured readers that India could be clearly divided into two ethnic groups – the more 'advanced' Aryan, or 'immigrant' group and the 'Tamulian', or 'aboriginal' group. This latter group, Hodgson proclaimed, was one which was, '... never distinguished by mental culture'. Brian Houghton Hodgson, 'A Brief Note on Indian Ethnology', *Journal of the Asiatic Society* 18 (1849).
95. Hodgson, 'A Brief Note on Indian Ethnology', 243.
96. Brian Houghton Hodgson, 'On the Kocch, Bodo, and Dhimal Tribes', *Miscellaneous Essays Relating to Indian Subjects* (London: Trubner & Co, 1880), 150.
97. Harrison, '"The Tender Frame of Man"', 85. Among other things, Martin would go on to sit on the committee appointed to investigate the sanitary state of the army in India which was ordered in 1859.
98. James Ranald Martin, *The Influence of Tropical Climates on European Constitutions, Including Practical Observations on the Nature and Treatment of the Diseases of Europeans on Their Return from Tropical Climates* (London: Churchill, 1856), 139.
99. See, for example, W. Falconer, *Remarks on the Influence of Climate, Situation, Nature of Country, Population, Nature of Food, and Way of Life, on the Disposition and Temper, Manners and Behaviour, Intellects, Laws and Customs, Forms of Government, and Religion, of Mankind* (London: C. Dilly, 1781), v.
100. Johnson, *The Influence of Tropical Climates, More Especially the Climate of India, on European Constitutions*, 433.
101. According to Ayurvedic practice, foods are divided into three main categories – *Rajsik, Tamsik* and *Satvik. Rajsik* foots (as the name implies) are foods associated with excess – or what a king might traditionally eat – and are heavy and rich. These foods are considered to be stimulants, sparking both sexual desire and aggression. *Tamsik* often denotes non-vegetarian foods or that prepared with many spices. These are considered to sap mental and bodily energies. Finally, *Satvik* food, usually vegetarian, is that prepared with the least amount of spices and is considered to be purifying.
102. C. A. Bayly, *Origins of Nationality in South Asia: Patriotism and Ethical Government in the Making of Modern India* (Oxford: Oxford University Press, 1998), 17–18. For a more detailed examination of Ayurvedic theories, see Francis Zimmerman, *The Jungle and the Aroma of Meats: An Ecological Theme in Hindu Medicine* (Berkeley: University of California Press, 1988).
103. The systematic emasculation of Indian men in general and Bengali men in particular has been examined in detail by John Rosselli and Mrinalini Sinha. See Sinha, *Colonial Masculinity*; John Rosselli, 'The Self-Image of Effeteness: Physical Education and Nationalism in Nineteenth-Century Bengal', *Past and Present* 86 (1980).
104. Martin, *The Influence of Tropical Climates on European Constitutions*, 139.
105. 'On the Medical Topography of Calcutta', *Calcutta Medical and Physical Society Quarterly Journal* 4 (1837), 649; ibid.
106. Kenneth Mackinnon, *A Treatise on the Public Health, Hygiene and Prevailing Diseases of Bengal and the North-West Provinces* (Cawnpore: Cawnpore Press, 1848), 8.

107. Dr Archibald Shanks, 'Report of H.M. 55th Foot', *The Madras Quarterly Medical Journal* 1, 2 (1839), 243.
108. Circular letter to the Superintending Surgeons of Divisions, 14 October 1831. NAI Home (Medical).
109. Ibid.
110. Medical Board Proceedings, January–April 1832. NAI Home (Medical).
111. Letter to Superintending Surgeon Langstaff, Meerutt from Assistant Surgeon MacGaveston, 3 December 1831. Medical Board Proceedings, no. 25 1832. NAI Home (Medical).
112. Ibid.
113. Ibid.
114. Letter to Superintending Surgeon Langstaff, Meerutt from Assistant Surgeon Finch, 12 January 1832. Medical Board Proceedings, no. 30 1832. NAI Home (Medical).
115. Letter to Superintending Surgeon Swiney, Kurnaul from Assistant Surgeon Grierson, 23rd Regiment Native Infantry, no. 21, 16 February 1832. Bengal Medical Board Proceedings, 8 March 1832. NAI Home (Medical).
116. Letter to Superintending Surgeon Langstaff, Meerutt from Assistant Surgeon Finch, 12 January 1832. NAI Home (Medical).
117. Letter to Superintending Surgeon Swiney, Kurnaul, from Assistant Surgeon George Smith, 37th Regiment, 22 February 1832. Bengal Medical Board Proceedings, 8 March 1832. NAI Home (Medical).
118. Letter to Superintending Surgeon Langstaff, Meerutt from Assistant Surgeon Palsgrave, Barielly, 17 December 1831. Bengal Medical Board Proceedings, 19 January 1832. NAI Home (Medical).
119. Letter to Superintending Surgeon Langstaff, Meerutt from Assistant Surgeon MacRae, 29 December 1831. Bengal Medical Board Proceedings, no. 25, 19 January 1832. NAI Home (Medical); Letter to Superintending Surgeon Langstaff, Meerutt from Assistant Surgeon MacGaveston, 3 December 1831. NAI Home (Medical); Letter to Superintending Surgeon Langstaff, Meerutt from Assistant Surgeon Finch, 12 January 1832. NAI Home (Medical).
120. George Ballingall, *A Probationary Essay on Syphilis: Submitted to the Royal College of Surgeons of Edinburgh* (Edinburgh: Balfour and Clarke, 1820), 2.
121. The dried root of a ground-dwelling orchid.
122. Letter to Superintending Surgeon Langstaff from Assistant Surgeon Bell, Meerutt, 29 December 1831. Medical Board Proceedings, no. 25, 19 January 1832. NAI Home (Medical); Letter to Superintending Surgeon Langstaff, Meerutt from Assistant Surgeon Finch, 12 January 1832. NAI Home (Medical); Letter to Superintending Surgeon Swiney, Kurnaul from Surgeon Thomson, 2nd Light Cavalry, 27 March 1832. Bengal Medical Board Proceedings, 9 April 1832. NAI Home (Medical).
123. Letter to Superintending Surgeon Langstaff, Meerutt from Assistant Surgeon MacRae, 29 December 1831. NAI Home (Medical); Letter to Superintending Surgeon Langstaff, Meerutt from Assistant Surgeon Finch, 12 January 1832. NAI Home (Medical).
124. Letter to Superintending Surgeon Playfair, Benares, from Surgeon Robert Tytler, 50th Regiment Native Infantry, no. 26, 29 November 1831. Bengal Medical Board Proceedings, 8 December 1831. NAI Home (Medical).

125. 'Wallace's Lectures on Syphilitic Eruptions', *Calcutta Medical and Physical Society Quarterly Journal* 2 (1837), 201.
126. G. A. Cowper, 'Identity of Syphilis and Gonorrhoea', *Calcutta Medical and Physical Society Quarterly Journal*, 2 (1837), 207.
127. Ibid., 208. These classifications hark back to Greek (and indeed unanic and ayurvedic) ways of identifying human 'types'.
128. 'Index', *Calcutta Medical and Physical Society Quarterly Journal* 1 (1837).
129. John Clark, 'Report on Syphilis in HM 13th Light Dragoons', *The Madras Quarterly Medical Journal* 1, 2 (1839), 384.
130. Ibid.
131. Benjamin Travers, *Observations on the Pathology of Venereal Affections* (London: Longman, Rees, Orme, Brown, and Green, 1830), 7.
132. Frederic Carpenter Skey, *A Practical Treatise on the Venereal Disease. Founded on Six Lectures on that Subject, Delivered in the Session of 1838–1839, at the Aldersgate School of Medicine. With Plates* (London: John Churchill, 1840), 4.
133. Clark, 'Report on Syphilis in HM 13th Light Dragoons', 386.
134. Ibid., 389. My italics.
135. John Murray, 'Quarterly Reports on the Health of Her Majesty's Troops in the Madras Command for 1838, by the Deputy Inspector General of Hospitals', *The Madras Quarterly Medical Journal* 1, 2 (1839), 36.

4 The body of the soldier and space of the cantonment

1. Stanley, *White Mutiny*, esp. 7–21. However, Stanley goes on to note that the Crown officers considered their Company peers to be inferior. See ibid., 37–8.
2. This chapter focuses on alcohol, however, there is certainly a need for further research on the military's attitude towards the use of opium and ganja among European troops in India.
3. See, most notably Foucault's 'Docile Bodies' chapter in Michel Foucault, *Discipline and Punish: The Birth of the Prison* (New York: Vintage Books, 1979).
4. Norman Chevers, 'On the Means of Preserving the Health of European Soldiers in India', *Indian Annals of Medical Science* 5 (1858), 748. In 1861, Chevers became the Secretary to the Bengal Medical Board and later served as the Principal of the Calcutta Medical College.
5. 'The European Soldier in India', *Calcutta Review* 59 (1858), 136.
6. Delirium tremens is a severe form of alcohol withdrawal perhaps best identified with the violent shaking fits that often accompany episodes. Literally translated, it means 'shaking madness'.
7. Chevers, 'On the Means of Preserving the Health of European Soldiers', 238.
8. 'The European Soldier in India', 136.
9. Present-day Kheda.
10. Gibson is typical of the kind of man who participated in the newly emerging medical and scientific societies across India. A vaccinator in the Deccan, he sent in a number of reports to the society sharing his observations and thoughts on salubrity in India.
11. Proceedings, Medical and Physical Society of Bombay, June–October. 1836, 49. MSA Publications, no. 1353.
12. 'Soldiers; Their Morality and Mortality', *Bombay Quarterly Review*, July (1855).

13. Ibid., 191.
14. Ibid., 175.
15. Albeit one which caused a lower level of troop admissions to hospital than venereal disease.
16. A bed or cot.
17. Macmullen, *Camp and Barrack Room*, 75.
18. Ibid.
19. Duke of Wellington Arthur Wellesley, 'Memorandum on the Proposed Plan for Altering the Discipline of the Army, 22 April 1829', *Despatches, Correspondence, and Memoranda, vol. 5* (London: John Murray, 1867), 592.
20. Ibid.
21. 'Soldiers; Their Morality and Mortality', 172.
22. Tellingly, despite the actual differences in the composition of Company versus Crown troops, the author made no distinction between the two.
23. Johnson, *The Influence of Tropical Climates, More Especially the Climate of India, On European Constitutions*, 401.
24. 'AD 1810, Regulation XX – A Regulation for Subjecting Persons Attached to the Military Establishment to Martial Law in Certain Cases, and for the Better Government of the Retainers and Dependants of the Army Receiving Public Pay on Fixed Establishments and of Persons Seeking a Livelihood by Supplying the Troops in Garrison, Cantonment and Station Military Bazars, or Attached to Bazars or Corps', *The Regulations of the Government of Fort William in Bengal, in Force at the End of 1853; to Which Are Added, the Acts of the Government of India in Force in That Presidency*, ed. Richard Clarke (London: J & H Cox, 1854), 6.
25. Macmullen, *Camp and Barrack Room*, 139. Emphasis mine.
26. Chevers, 'On the Means of Preserving the Health of European Soldiers'.
27. Ibid., 679.
28. Ibid.
29. William Howard Russell, *My Indian Mutiny Diary*, ed. Michael Edwardes (London: Cassell & Company, 1860), 100–1.
30. Suleri, *The Rhetoric of British India*, 31.
31. Foucault, *Discipline and Punish*, 135.
32. Harrison, *Climates and Constitutions*, 58.
33. Concerns about over-eating in hot climates, though not as great perhaps as those about drink, continued throughout much of the century. Such a concern could explain the (unproven) story of the death of Rose Aylmer (herself the subject of a poem by Walter Savage Landour), buried in Calcutta's South Park Street Cemetery who is said to have died, at the age of 20, of a pineapple overdose. There was certainly a belief that, as one British soldier wrote to his father, '... fruit eating is very injurious in this country'. William Porter, Letter to James Porter from William Porter, Penang, 14 August 1828. Porter Papers APAC MSS EUR G128.
34. Johnson, *The Influence of Tropical Climates, More Especially the Climate of India, on European Constitutions*, 68.
35. Ibid., 138.
36. Letter to Lieutenant Colonel J. Stuart, Officiating Secretary to the Government of India, Military Department from the Judge Advocate General, Simla 10 June 1839. Consultations, nos. 15–19, 8 July 1839. NAI Home (Military).

37. Ibid., 91.
38. Medical Topographical Report of the Military Stations occupied by His Majesty's troops in the Presidency of Bengal for the year 1827. Medical Topographical Reports, 1827–1860 NAI Military (Miscellaneous).
39. Letter to Lieutenant Colonel J. Stuart, Officiating Secretary to the Government of India, Military Department from the Judge Advocate General, Simla 10 June 1839. NAI Home (Military).
40. Mr Surgeon Ainslie's Plan for Preserving the Health of the European Soldiers, Deemed It Inexpedient to Adopt, 1807 Board of Commissioners Collections 1808–1809. APAC F/4/226/4983.
41. Harrison, *Climates and Constitutions*, 91.
42. Letter to Elizabeth Eggleston from Gunner William Hurd Egglestone, Horse Brigade Artillery, St Thomas Mount, 10 August 1846. APAC MSS EUR PHOTO EUR 257.
43. Chevers, 'On the Means of Preserving the Health of European Soldiers'.
44. Marc Jason Gilbert, 'Empire and Excise: Drugs and Drink Revenue and the Fate of States in South Asia', *Drugs and Empires: Essays in Modern Imperialism and Intoxication, c. 1500–1930*, 116–41 eds. James Mills and Patricia Barton (Basingstoke: Palgrave Macmillan, 2007), 117.
45. The volume of a dram was an inexact measure, but is roughly equivalent to 30ml.
46. Sir Monier Monier-Williams, *A Few Remarks on the Use of Spirituous Liquors among the European Soldiers; and on the Punishment of Flogging in the Native Army of the Honourable the East India Company* (London: DS Maurice, 1823), 6.
47. Monier Williams, *A Few Remarks on the Use of Spirituous Liquors among the European Soldiers*, 8.
48. Henry Piddington, *A Letter to the European Soldiers in India, on the Substitution of Coffee for Spirituous Liquors* (Calcutta: The Englishman Press, 1839).
49. 'AD 1809, Regulation III – A Regulation for the Support of the Police in the Cantonments and Military Bazars, for Defining the Powers of the Civil and Military Officers in the Performance of that Duty; and for Fixing the Local Limits of the said Cantonments and Bazars', *The Regulations of the Government of Fort William in Bengal, in Force at the End of 1853; to Which are Added, the Acts of the Government of India in Force in that Presidency*, ed. Richard Clarke (London: J & H Cox, 1854), 3.
50. Ibid.
51. Granted, Havelock was a tee-total Baptist who, before his participation in breaking the siege of Lucknow (and death from dysentery!) made him infamous, organised the distribution of Bibles for soldiers as well as all-rank Bible study groups.
52. Present-day Jalalabad.
53. Henry Havelock, *Memoirs of Major General Sir Henry Havelock*, ed. John Clark Marshman (London: Longman, Green, Longman and Roberts, 1860), 109.
54. Peers, 'Imperial Vice', 45. Chima Korieh has similarly argued the importance of the alcohol monopoly to the imperial state in the late nineteenth and early twentieth century Nigeria, where much of the official revenue was generated from liquor tariffs and duties. See Chima J. Korieh, 'Dangerous Drinks and the Colonial State: "Illicit" Gin Prohibition and Control in Colonial

Nigeria', *Drugs and Empires: Essays in Modern Imperialism and Intoxication, c. 1500–1930*, 101–15 eds. James Mills and Patricia Barton (Basingstoke: Palgrave Macmillan, 2007), 103.
55. Chevers, 'On the Means of Preserving the Health of European Soldiers', 689.
56. Ibid. These plant-based drugs induced a range of reactions, from hallucinations to paralysis.
57. Ibid., 688.
58. Ibid., 690.
59. Ibid.
60. For a Foucaultian analysis of disciplinary measures in late colonial Delhi, see Stephen Legg, 'Governing Prostitution in Colonial Delhi: From Cantonment Regulations to International Hygiene (1864–1939)', 34, 4 (2009); Stephen Legg, *Spaces of Colonialism: Delhi's Urban Gvernmentalities* (Oxford: Blackwell, 2007).
61. 'AD 1809, Regulation III – A Regulation for the Support of the Police in the Cantonments and Military Bazars, for Defining the Powers of the Civil and Military Officers in the Performance of that Duty; and for Fixing the Local Limits of the said Cantonments and Bazars', 97.
62. Nicholas B. Dirks, *The Scandal of Empire: India and the Creation of Imperial Britain* (Cambridge, MA: The Belknap Press of Harvard University Press, 2006), 170.
63. 'AD 1810, Regulation XX – A Regulation for Subjecting Persons Attached to the Military Establishment to Martial Law in Certain Cases, and for the Better Government of the Retainers and Dependants of the Army Receiving Public Pay on Fixed Establishments and of Persons Seeking a Livelihood by Supplying the Troops in Garrison, Cantonment and Station Military Bazars, or Attached to Bazars or Corps', 154.
64. The Articles of War are the body of regulations that compose military law.
65. 'AD 1810, Regulation XX, Section II.
66. Ibid.
67. Letter to Mr Chief Secretary Hill from W. Morison, Commissary General, 4 October 1826. Enactment of Regulation 7 of 1832 for the Better Discipline of Military Bazars and Relative to the Establishments to Be Entertained in Lieu of Troops at the Several Military Stations for the Execution of Civil Process. Board of Commissioners Collections 1833–1834. APAC F/4/1425/56238.
68. Ibid.
69. 'AD 1810, Regulation XX, Section VI.
70. Ibid., Section XXVI.
71. T. Jacob, *Cantonments in India: Evolution and Growth* (New Delhi: Reliance Publishing House, 1994), 113.
72. Regulation VII of 1832; A Regulation for [...] the better order and discipline of Military Bazars, the more effective administration of justice, and of the police, at the stations where such Bazars are established, and at certain other military stations, and in military forces in the field; the extension of the powers of Courts Martial; and the more effectual prevention of the undue use of spirituous and fermented liquors, and intoxicating drugs, by

the European Troops under this Presidency. Board's Collections 1833–1834. APAC F/4/1425/56238.
73. Ibid.
74. Ibid.
75. Ibid.
76. A *pandal* is a temporary structure for religious purposes, often associated with veneration for the Goddess Durga during the Durga Puja.
77. The *pettah* is the town which surrounds the fort or cantonment.
78. This weight conversion varied, but was roughly equivalent to 2 ½ lbs.
79. Award of Contract to Sell Liquor in the Bazaar to Luxemon Mahodjee Roomdar. 9 February 1825, 2–3. MSA Military Department.
80. Ibid.
81. Number 179 General Order by the Governor General in Council, 12 September 1836, cited in, T. Jacob, *Cantonments in India: Evolution and Growth* (New Delhi: Reliance Publishing House, 1994), 115.
82. Ibid. My italics.
83. For vagrancy in colonial India more generally, see Arnold, 'European Orphans and Vagrants in India'.
84. Letter from H. Havelock, Adjutant General, 3 September 1847. Bombay Military Proceedings, 22 September 1847. APAC P/363/36
85. Proceedings of a Committee Assembled by Order of Brigadier Feron, Commanding at Poona, to Report on the Quality of Beer Brewed at Poona, 19 March 1839. Proceeding 34, Regarding Brewing of Beer 1839. MSA Military Department Compilation, vol. 119.
86. Brewers often shipped beer from England to India. The most famous of these exports, India Pale Ale, continues to be produced today. For a brief history of the development of India Pale Ale, see Alan Pryor, 'Indian Pale Ale: An Icon of Empire', *Commodities of Empire, Working Paper Series* 13 (2009).
87. Letter to the Military Board from the Honourable Court of Directors, no. 14, dated 31 July 1840. Medical Board Proceedings, 9 November 1840. NAI Home (Medical).
88. Ibid.
89. Hugh Macpherson, 'Analysis of the Later Medical Returns of European Troops Serving in the Bengal Presidency', *Indian Annals of Medical Science* 5 (1858), 238.
90. 'The European Soldier in India', 136.
91. F. S. Arnott, 'Report on the Health of the 1st Bombay European Regiment (Fusiliers), from 1st April 1846 to 31st March 1854', *Transactions of the Medical and Physical Society of Bombay* 2 (1855), 110.
92. Ibid., 112.
93. Ibid., 112–13.
94. Letter to Charles Compton from Richard Compton, 12 Royal Lancers, Bangalore, dated 21 February 1857. APAC MSS EUR C243.
95. Sharon Murphy, 'Libraries, Schoolrooms, and Mud Gadowns: Formal Scenes of Reading at East India Company Stations in India, c. 1819–1835', *Journal of the Royal Asiatic Society* Series 3, 21, 4 (2011), 465.
96. This rule was later relaxed and the men were allowed to take books into the barracks with them (a fact which enabled the literate among them to read aloud to their non-literate peers). See ibid. Murphy does not state the percentage of literate soldiers, making it difficult to assess the potential success of these reading rooms.

97. *Rules Establishing Regimental Savings Banks in the Regiments of the British and Indian Armies Serving in Bengal, with the Forms in Use* (Calcutta: Military Orphan Press, 1860), 4.
98. Chevers, 'On the Means of Preserving the Health of European Soldiers', 760.
99. Number 257 of 1837. General Order by His Excellency Commander-in-Chief, Headquarters Simla, 16 January 1838, by the Honourable President in Council, Fort William, 29 December 1837. General Orders Issued to Honourable Company's Troops on the Bengal Establishment 1838. NAI Military (Miscellaneous).
100. Arnott, 'Report on the Health of the 1st Bombay Fusiliers', 139.
101. Rules for Guidance of Soldiers in Regard to Mode of Living – Sanitary Rules for the Use of European Troops. 26 July and 23 August 1858. NAI Home (Medical).
102. Ibid.
103. For gender and the symbolic and emotive role of (albeit supposedly 'upright') European women in India see Douglas Peers, '"The more this foul case is stirred, the more offensive it becomes": Imperial Authority, Victorian Sentimentality and the Court Martial of Colonel Crawley, 1862–1864', *Fringes of Empire: Peoples, Places and Spaces in Colonial India*, 207–35 eds. Sameeta Agha and Elizabeth Kolsky (Delhi: Oxford University Press, 2009). For the 'domestication' of empire, see Anna Davin, 'Imperialism and Motherhood', *History Workshop Journal* 5 (1978).
104. Stoler, 'Rethinking Colonial Categories', 139.
105. Letter to G. J. Casamajor Esq, Acting Secretary to the Government from G. A. Weatherall, Secretary to Commander-in-Chief, Fort St George, 15 November 1824. Question as to the Allowances of the Wives Widows and Children of European Soldiers. Madras Military Collections 1825. APAC F/4/787/21363.
106. Minute from Thomas Munro, November 1824. APAC F/4/787/21363.
107. Ibid. Of course, Munro's daughter, Jessie Thompson, was born of an Indian mother. See Burton Stein, *Thomas Munro: The Origins of the Colonial State and His Vision of Empire* (Delhi: Oxford University Press, 1989), 247.
108. Minute from Thomas Munro, Governor in Council, November 1824. Military Secretary's Office, Madras Military Collections. 1825. APAC F/4/787/21363.
109. 'Soldiers; Their Morality and Mortality', 203.
110. Ibid., 204. Doll Tearsheet is the fictional favourite prostitute of Falstaff in Shakespeare's *Henry IV*.
111. Ibid.
112. Minute from Lieutenant Colonel H. Havelock, Deputy Adjutant General HM Forces Bombay, 23 May 1848. Bombay Military Consultations, 24 May 1848. APAC P/363/49/2245.
113. Ibid.
114. General Orders Issued from the Adjutant General's Office. 1 July to 31 December 1813, vol. 28, Madras 1813. APAC L/MIL/17/3/362.
115. Letter to the Secretary to Government in the Public Department from the Superintendent of Police, 20 December 1822. Madras Public Proceedings, 22 October to 31 December 1822. APAC P/245/37.

116. Ibid.
117. Letter to Captain Ormsby from Mary Ann MacMullen, 8 March 1823. TNSA Fort St George Public Consultations, 25 March 1823, vol. 506; Letter to Captain Ormsby from Eliza Williams (n.d.). TNSA Fort St George Public Consultations, 25 March 1823, vol. 506.
118. Letter to Lieutenant Colonel Blair, Military Secretary from Lieutenant Nagel of HM 47th Regiment, Poona. 14 September 1824. MSA Military Department, vol. 114.
119. This was the case for two women of the 20th Foot and Queens Royals who had been ordered home on the *Lonach* but fled the depot into Bombay shortly before the ship's departure and could not be found. See Letter to the Senior Magistrate of Police from Chief Secretary, Military Department, Bombay. 13 February 1826. MSA Military Department, vol. 4.
120. Letter to the Honourable the Court of Directors from Military Department, 30 October 1822. Bombay Military Letters Received, 31 October 1822 to 24 December 1824. APAC L/MIL/3/1724.
121. Judge Advocate General's Opinion in the case of Mrs Philips. Measures to be Adopted for Dealing with European Women whose Habitual Misconduct may Require their Removal from Regimental Lines 1865. MSA Military Department (1864–1868), vol. 1115, no. 291 of 1865.
122. Letter to Adjutant Royal Artillery from [no sender given], Ahmedabad 14 February 1868. Measures to be Adopted for Dealing with European Women whose Habitual Misconduct may Require their Removal from Regimental Lines 1865. MSA Military Department (1864–1868), vol. 1115, no. 291
123. Letter to the Secretary to Government, Military Department from the Quarter Master General of the Army, Fort St George 20 February 1862. Military Consultations, nos. 321–22, 17 March 1862. TNSA Madras Government Proceedings.
124. Letter to His Excellency the Right Honourable the Governor General of India in Council from C. Wood, Military Department, Simla 20 May 1865. Measures to be Adopted for Dealing with European Women whose Habitual Misconduct may Require their Removal from Regimental Lines 1865. MSA Military Department (1864–1868), vol. 1115, no. 291.
125. Letter to Lieutenant Colonel Stuart, Officiating Secretary to the Government of India in the Military Department from Major Cragie, Deputy Adjutant General, 22 April. Proceedings, no. 32, 28 May 1842. NAI Home (Military).
126. For reference to other forms of punishment, see Letter to Lieutenant Colonel J. Stuart, Officiating Secretary to the Government of India, Military Department from the Judge Advocate General, Simla 10 June 1839. NAI Home (Military).
127. While it appears that these character books were kept in most regiments, these books have either been lost or destroyed and only secondary references to them still exist. See, for example, the court martial of Gunner William Carter, in which an extract of Carter's 'Character Book' is included. See Extract from the Character Book of the 1st Company, 1st Battalion

Artillery, Agra, 28 February 1845. Proceedings, no. 82, 28 March 1845. NAI Home (Military).

128. Monier-Williams, *A Few Remarks on the Use of Spirituous Liquors*, 6.
129. Extract from Proceedings of a General Court Martial assembled at Cawnpore, on Wednesday, 5 March 1845 for the Trial of Lieutenant James Goodlad Wollen, of the 42nd Regiment of Light Infantry. Proceedings, no. 226, 27 June 1845. NAI Home (Military).
130. Petition to the Queen's Most Excellent Majesty, the Humble Petition of Jane Taylor widow of the late Sergeant Taylor of the East India Company's Service, 24 April 1854. Proceedings, nos. 419–20, 5 May 1854. NAI Home (Military).
131. Extract of Proceedings of a General Court Martial held for the trial of Assistant Surgeon Alexander Storm of the 51st Regiment, Native Infantry. Proceedings, no. 127, 13 April 1835. NAI Home (Military).
132. Ibid.
133. Extract from the Character Book of the 1st Company, 1st Battalion Artillery, Agra, 28 February 1845. NAI Home (Military).
134. 'Soldiers; Their Morality and Mortality', 196.
135. Cashiering is the (often very visible and public) process of dishonourable dismissal from service.
136. Extract from proceedings of a General Court Martial assembled at Sukkur for the Trial of Lieutenant and Brevet Captain George Becher, 4th Regiment Bengal Native Infantry, 15 December 1844. Proceedings, no. 210, 25 July 1845. NAI Home (Military).
137. Extract from Proceedings of a General Court Martial held at Dinapore for the trial of Ensign Alexander MacNeill of the 72nd Regiment Native Infantry, March and April 1845. Proceedings, no. 218, 25 July 1845. NAI Home (Military).
138. Ibid.
139. I am grateful to Doug Peers for this reminder. As the sessions were 'closed', no records were released, making the archival trail even more difficult to trace.
140. William Hough, *The Practice of Courts-Martial, also the Legal Exposition and Military Explanation of the Military Act, and Articles of War, Together with the Crimes and Sentences of Numerous Courts-Martial, and the Remarks Thereupon by His Majesty, and the Several Commanders in-Chief in the East Indies, and on Foreign Stations, &c., the Whole Forming a Manual of the Judicial and Military Duties of an Officer in Various Situations* (London 1825), 819–20.
141. Indeed, a 1794 Calcutta rape case illuminates this point perfectly. The case involves a woman named Peerun (identified as the 'wife of Bukshu, a Khedmetgar') who was assaulted and raped by Iman Buksh. Peerun (who was six months pregnant at the time) was walking along the Esplanade with four of her friends when they were approached by Buksh alleging to be a chuprassy and threatening to take them to the thana if they didn't give him money. He beat away Peerun's four friends, but held her back, dragging her down and propositioning her for sex. She begged him to stop, protesting that she was pregnant, but he gagged her mouth with a piece of cloth and forced himself on her. The court asked Peerun if she felt 'anything

wet come from him'. She answered that she did not know. The fact that 'emission was not proved' and that she 'had not made such resistance' (while she tried to keep her legs closed, he eventually forced them open), convinced the court that the incident did not constitute a rape and Buksh was found 'not guilty'. See *Case of Peerun, a Woman, against Iman Buksh, for a Rape, 9 December 1794. High Court Bar Library, Terms 1 & 2, 1794*, 1794. Victoria Memorial Library, Hyde Papers, Kolkata.

142. Hough, *The Practice of Courts-Martial*, 705.
143. Rules and Articles for the Better Government of the Officers and Soldiers in the Service of the East India Company from 1 January 1858. APAC L/MIL/5/422/382.
144. Ibid.
145. Ibid., and *Rules Establishing Regimental Savings Banks in the Regiments of the British and Indian Armies Serving in Bengal, with the Forms in Use*
146. Sinha, *Colonial Masculinity*.
147. As in the case of Rebecca Lane, who was accused of attempting to poison her husband, along with her lover, Jonathan Taylor. See Petition to the Queen's Most Excellent Majesty, the Humble Petition of Jane Taylor widow of the late Sergeant Taylor of the East India Company's Service, 24 April 1854. NAI Home (Military).
148. Letter to Lieutenant Colonel R. J. H. Birch, Officiating Secretary to the Government of India in the MD, from T. Caird Esq, Sheriff of Calcutta transmitting a Petition to the Most Noble Jo Andrew Marquis of Dalhousie, Governor General of India, the Humble Petition of Thomas Pacey, 3 May 1854. Proceedings, no. 424, 5 May 1854. NAI Home (Military). A report of the convicts arrived in Western Australia in 1855 suggests, however, that Pacey's petition was unsuccessful, as he arrived there with six other convicts on 9 January 1855. See 'Convicts to Australia Guide – Arrived in Western Australia, 1855', http://members.iinet.net.au/~perthdps/convicts/con-wa14.html.
149. Hough, *The Practice of Courts-Martial*, 665.
150. Ibid.

5 'Unofficial' responses to lock hospital closure, 1835–1868

1. Levine, 'Rereading the 1890s', 592.
2. Ibid.
3. Ibid., 593.
4. Ibid., 587.
5. Levine, *Prostitution, Race, and Politics*, 95.
6. Separate Military Letter from Bengal, 29 May 1832. APAC F/4/1338/53031.
7. Ibid.
8. Report on Measures Adopted for Sanitary Improvements in India, from June 1872 to June 1873. Memorandum by the Army Sanitary Commission on the Annual Report on the Cantonment Lock Hospitals in the Madras Presidency for the Year 1871. APAC V/24/3677.
9. Extracts from the Annual Medical Reports of Deputy Inspectors of HM Hospitals at Bombay for 1834, 1836 and 1837. Letter to the Medical Board from Lieutenant Colonel Stuart, Secretary to Government, Military

Department, 7 April 1841. Medical Board Proceedings, 15 April 1841. NAI Home (Medical).
10. Enclosure, Précis of Correspondence Regarding Lock Hospitals. Letter to the Secretary to Government, Bombay Military Department, from the Adjutant General of the Army, 11 April 1863. MSA Military Department (1863–1867), vol. 1133
11. M. Sundara Raj, *Prostitution in Madras: A Study in Historical Perspective* (Delhi: Konark, 1993), 31.
12. This is not to dismiss the 'pacification' campaigns which took place in the 1830s and 1840s, nor the fact that a number of Bombay regiments served in the Afghanistan campaigns, but both presidencies saw no direct risk to their borders at this time.
13. Extracts from the Annual Medical Reports of Deputy Inspectors of HM Hospitals at Bombay for 1834, 1836 and 1837. Letter to the Medical Board from Lieutenant Colonel Stuart, Secretary to Government, Military Department, 7 April 1841. NAI Home (Medical).
14. Letter to Colonel Casement from W. Burke, 21 April 1832. APAC F/4/1338/53031.
15. Extract of a Letter to Captain Keith, Acting Adjutant General from D. MacLeod, Deputy Inspector General of Hospitals, 24 October 1831. Correspondence with the Bombay Government Respecting the Abolition of Lock Hospitals at Bombay. Board of Commissioners Collections 1832–1833. APAC F/4/1338/53031.
16. Ibid.
17. Letter to Colonel Casement from W. Burke, 21 April 1832. APAC F/4/1338/53031.
18. Ibid., 80.
19. Ibid.
20. By 'public', Burke appears to be more narrowly referring to those individuals resident in the cantonment.
21. Circular to the Medical Officers in Charge of HM Regiments & Depot, from John Murray, Inspector General of Hospitals HM Troops, 11 June 1838. Madras Military Consultations, 21 August 1838. APAC P/267/10.
22. Ibid.
23. Letter to Major Fearon from Mr Murray, Inspector General of Hospitals, 14 July 1838. Madras Military Consultations, 21 August 1838. APAC P/267/10.
24. Lorenzo M. Crowell, 'Military Professionalism in a Colonial Context: The Madras Army, circa 1832', *Modern Asian Studies* 24, 2 (1990), 249.
25. Letter to John Murray, Deputy Inspector General of Hospitals, Madras from T. Servis, MD Surgeon, 4th King's Own, 14 June 1838. Madras Military Consultations, 21 August 1838. APAC P/267/10.
26. Ibid. My italics.
27. Summary Statement of the Comparative Prevalence of Venereal Disease in Her Majesty's Hospitals, Prior and Subsequent to the Abolition of the Lock Hospitals; Taken from Official Reports and Returns in the Office of the Deputy Inspector General of Hospitals. Madras Military Consultations, 21 August 1838. APAC P/267/10.
28. Letter to John Murray, Deputy Inspector General of Hospitals, Madras from T. Servis, MD Surgeon, 4th King's Own, 14 June 1838. APAC P/267/10. Here,

however, Servis does depart from the average lock hospital advocate in his belief that men, as well as women, should be subject to these examinations.

29. Letter to John Murray, Deputy Inspector General of Hospitals, Her Majesty's Forces, From J. Mouat, MD, Surgeon, HM 13th Dragoons, Bangalore 30 June 1838. Madras Military Consultations, 21 August 1838. APAC P/267/10.
30. Ibid.
31. The corpus of literature on Company and Crown responses to famine, and the broader implications for Indian society and understandings of empire, is fast-growing. See, for example, Mike Davis, *Late Victorian Holocausts: El Nino Famines and the Making of the Third World* (London: Verso, 2000); David Arnold, 'Hunger in the Garden of Plenty: The Bengal Famine of 1770', *Dreadful Visitations: Confronting Natural Catastrophe in the Age of Enlightenment*, ed. Alessa Johns (New York, London: Routledge, 1999), 81–112; David Hall-Matthews, *Peasants, Famine and the State in Colonial Western India* (Basingstoke: Palgrave Macmillan, 2005).
32. Letter to John Murray, Deputy Inspector General of Hospitals, Her Majesty's Forces, From J. Mouat, MD, Surgeon, HM 13th Dragoons, Bangalore 30 June 1838. APAC P/267/10. Underline in original.
33. Ibid.
34. Ibid.
35. Ibid.
36. Letter to the Quarter Master General of the Army from the Officer Commanding Mysore Division, 3 July 1855. Madras Military Consultations, 30 July to 26 August 1856. APAC P/273/41.
37. Levine, *Prostitution, Race, and Politics*, 88.
38. Letter to John Murray, Deputy Inspector General of Hospitals, Her Majesty's Forces, From J. Mouat, MD, Surgeon, HM 13th Dragoons, Bangalore 30 June 1838. APAC P/267/10.
39. Enclosure, Précis of Correspondence Regarding Lock Hospitals. Letter to the Secretary to Government, Bombay Military Department, from the Adjutant General of the Army. MSA Military Department (1863–1867), vol. 1133.
40. Report of the Medical Board. Madras Military Consultations, 21 August 1838. APAC P/267/10.
41. Letter to the Secretary to the Government of India, Military Department from Mr Steel, Secretary to Government, Fort St George 6 November 1838. Consultations, nos. 70–72, 4 February 1839. NAI Home (Military).
42. Extract from the Proceedings of the Most Noble the Governor General of India in Council in the Home Department, 10 February 1852. Military Consultations, nos. 174–5, 27 February 1852. NAI Home (Military).
43. Hazaribagh.
44. West Bengal State Archives (hereafter WBSA) *Letter to Major ET Dalton, Commissioner of Chota Nagpore, from T. Simpson, Deputy Commissioner of Hazareebagh, dated 10 November 1861*. Judicial Proceedings, no 169, November 1861. WBSA Judicial.
45. Ibid.
46. Letter to Lieutenant Colonel Browne, Secretary to Government in the Military Department, from Lieutenant Colonel Alexander, Adjutant General of the Army, 1 March 1849. Measures Taken with the View of Mitigating the Prevalence of Venereal Disease Among the European Troops. Madras Military. Board of Commissioners Collections 1849–1850. APAC F/4/2341.

47. Letter to Adjutant General from Captain T. R. Morse, Superintendent Bazars & Police Poona, 16 September 1848. Bombay Military Consultations, 8 November 1848. APAC P/363/56.
48. Ibid.
49. Minute by the Military Board, 19 October 1848. Bombay Military Consultations, 8 November 1848. APAC P/363/56.
50. Enclosure, Précis of Correspondence Regarding Lock Hospitals. Letter to the Secretary to Government, Bombay Military Department, from the Adjutant General of the Army. MSA Military Department (1863–1867), vol. 1133.
51. Ibid.
52. Letter to the Secretary to Government, Military Department from the Quarter Master General of the Army, Fort St George 20 February 1862. TNSA Madras Government Proceedings.
53. Chatterjee, *The Nation and Its Fragments*, 6.
54. Judy Whitehead, 'Bodies Clean and Unclean: Prostitution, Sanitary Legislation, and Respectable Femininity in Colonial North India', *Gender and History* 7, 1 (1995), 51.
55. Letter to Medical Board from Lieutenant Colonel Stuart, Secretary to Government in Military Department, no. 17, 7 April 1841. Medical Board Proceedings, 15 April 1841. NAI Home (Medical).
56. A petition drafted by unnamed 'common prostitutes' in Calcutta in 1868 (alleging mistreatment by the police), suggests the use of dispensaries for venereal treatment was, by that time, common practice. See *Prostitute's Registration Ticket*. Proceedings, no. 67, April 1869. WBSA General (Sanitation).
57. Ramanna, *Western Medicine and Public Health in Colonial Bombay*, 51.
58. Ibid., 54.
59. Report of the Medical Board. APAC P/267/10.
60. Ibid.
61. Letter to the Adjutant General of the Army, from Brigadier J. L. James, 17 June 1848. Measures Taken with the View of Mitigating the Prevalence of Venereal Disease among the European Troops. Madras Military. Board of Commissioners Collections 1849–1850. APAC F/4/2341.
62. Ibid. The order did not specify a type of punishment. The system operated for four years, until 1848, when Brigadier James moved to stop it, noting the corruption and extortion which it promoted in the cantonment.
63. Register of In- and Out-Patients Admitted to the Benefit of the Native Hospital at Benares, 1 January to 31 December 1840. Medical Board Proceedings, 11 March 1841. NAI Home (Medical).
64. Letter to the Calcutta Medical Board, from the Secretary to Government North West Provinces, 27 February 1841. Medical Board Proceedings, 11 March 1841. NAI Home (Medical).
65. Half Yearly Reports of the Government Charitable Dispensaries, Established in the Bengal and North Western Provinces, from 1 February to 31 July 1842; 1 August 1842 to 31 January 1843; and 1 February to 31 July 1843. APAC V/24/733.
66. Sittwe, Burma. Index for 23 October 1843. Medical Board Proceedings 1843. NAI Home (Medical).
67. Letter from Dr Clapperton, forwarding Half Yearly Returns of House- and Out-Patients Treated at the Government Dispensary at Moorshidabad, 25 August 1843. Medical Board Proceedings, 28 August 1843. NAI Home (Medical).

68. Letter to the Medical Board from Mr Steven, 17 February 1845; Forwarding Half Yearly Reports of House- and Out-Patients Treated in the Agra and Muttra Government Dispensaries, Ending 31 January 1845; Solicits the Board's Attention to the Increase of Syphilitic Disease in the City of Agra. Medical Board Proceedings, 27 February 1845. NAI Home (Medical).
69. Letter to Adjutant General from Captain T. R. Morse, 16 September 1848. APAC P/363/56.
70. Enclosure, Précis of Correspondence Regarding Lock Hospitals. Letter to the Secretary to Government, Bombay Military Department, from the Adjutant General of the Army. MSA Military Department (1863–1867), vol. 1133.
71. Ibid.
72. Letter to J. F. Thomas, Chief Secretary to Government, from Assistant Surgeon A. Lorimer, Secretary Medical Board, 5 July 1849. Measures Taken with the View of Mitigating the Prevalence of Venereal Disease among the European Troops. Madras Military. Board of Commissioners Collections 1849–1850. APAC F/4/2341.
73. Reports on the Venereal Dispensaries at Trichinopoly and Bellary. Madras Public Consultations, 4 June to 23 July 1844. APAC P/248/13.
74. Letter to J. F. Thomas from Assistant Surgeon A. Lorimer, 5 July 1849. APAC F/4/2341.
75. Reports on the Venereal Dispensaries at Trichinopoly and Bellary. APAC P/248/13.
76. Ibid.
77. It is unfortunately not within the scope of this book to discuss public health, but only to suggest the places where military concerns intersected, and clashed, with nascent 'public' ones.
78. For a study of mesmerism in India, and Esdalie in particular, see Waltrud Ernst, 'Colonial Psychiatry, Magic and Religion. The Case of Mesmerism in British India', *History of Psychiatry* 15, 1 (2004).
79. Hugli.
80. *Half Yearly Reports of the Government Charitable Dispensaries Established in Bengal and North Western Province, from 1 April to 30 September* (Calcutta: Bengal Orphan Press, 1849).
81. Ibid.
82. Extract Public Letter to Fort St George, 28 April 1847. Measures Taken with the View of Mitigating the Prevalence of Venereal Disease among the European Troops. Madras Military. Board of Commissioners Collections 1849–1850. APAC F/4/2341.
83. Ibid.
84. Letter to the Major of the Brigade, Malabar and Canara, from Superintending Surgeon D. S. Young, 10 January 1849. Measures Taken with the View of Mitigating the Prevalence of Venereal Disease Among the European Troops. Madras Military. Board of Commissioners Collections 1849–1850. APAC F/4/2341.
85. The fact that the women were *asked* about their willingness to return home suggests that, unlike other cantonments, the local magistrate was unwilling to authorise the forced expulsion of unwanted women from the cantonment lines.
86. Chatterjee, 'Colouring Subalternity', 69–70; Ghosh, *Sex and the Family in Colonial India*, 18.

87. Letter to the Major of the Brigade Malabar and Canara from Superintending Surgeon D. S. Young, 10 January 1849. APAC F/4/2341.
88. Letter to the Secretary to the Medical Board, Madras, from Superintending Surgeon Young, 6 June 1849. Measures Taken with the View of Mitigating the Prevalence of Venereal Disease among the European Troops. Madras Military. Board of Commissioners Collections 1849–1850. APAC F/4/2341.
89. Ibid.
90. Minute of Government, no. 806. Measures Taken with the View of Mitigating the Prevalence of Venereal Disease among the European Troops. Madras Military. Board of Commissioners Collections 1849–1850. APAC F/4/2341.
91. Letter to Lieutenant Colonel Havelock, Deputy Adjutant General, from J. Kinnis, Deputy Inspector General HMs Hospitals, 9 May 1848. Minutes from the Deputy Adjutant General, 16 May 1848. Bombay Military Consultations, 15 July 1848. APAC P/363/49/2398.
92. Ibid.
93. Ibid.
94. Letter to the Secretary to Government, Military Department, from the Quarter Master General of the Army, 26 June 1855. Madras Military Consultations, 17 July 1855. APAC P/273/23.
95. Respectively, Kamthi, Poovirundavalli, St Thomas' Mount and Jalna. Letter to the Secretary to Government, Military Department from the Quarter Master General of the Army, Fort St George 20 February 1862. TNSA Madras Government Proceedings.
96. Memorandum from the Adjutant General of the Army, 12 February 1862. Proceedings, Madras Government, no. 732, 25 February 1862. TNSA Military Department.
97. *Royal Commission to Inquire into the Organization of the Indian Army* (London: HM Stationery Office, 1859).
98. Ibid., 258.
99. Ibid., 257.
100. *Report of the Commissioners Appointed to Inquire into the Sanitary State of the Army in India; with Précis of Evidence* (London: H.M. Stationery Office, 1863).
101. Ibid., 16.
102. 'Obituary. Norman Chevers, C.I.E., M.D., F.R.C.S. Eng, Deputy Surgeon General, HM Indian Army', *The British Medical Journal*, December (1886).
103. E. I. Carlyle, 'Tulloch, Sir Alexander Murray (1803–1864)', *Oxford Dictionary of National Biography*, ed. John Sweetman (London: Oxford University Press, 2004).
104. *Report of the Commissioners Appointed to Inquire into the Sanitary State of the Army in India; with Précis of Evidence* 24–27.
105. Ibid., 24.
106. Ibid., 26.
107. Ibid., 33.
108. Ibid., 38.
109. Ibid., 39.
110. The Report noted that sepoys in the Madras and Bombay armies were allowed to have their families with them, while Bengal sepoys used their furlough periods to visit their wives and children.

111. Q. 533–538, Testimony of Colonel Swatman, *Report of the Commissioners Appointed to Inquire into the Sanitary State of the Army in India; with Précis of Evidence* 46.
112. Ibid., 47.
113. Q. 2393, Testimony of Dr MacLean, ibid., 143.
114. Q. 2393, Testimony of Dr MacLean, ibid.
115. Ibid., 143. Testimony of Dr MacLean, no. 2393.
116. Ibid., 144. Testimony of Dr MacLean, nos. 2402–03.
117. Ibid., lxii.
118. Ibid. My italics.
119. *Memorandum on Measures Adopted for Sanitary Improvements in India up to the End of 1867; Together with Abstracts of the Sanitary Reports Hitherto Forwarded from Bengal, Madras and Bombay* (London: George Edward Eyre and William Spottiswoode, 1868), 70.
120. Minute by His Excellency the Governor. 22 April 1863. MSA Military Department (1863–1867), vol. 1133.
121. Minute by His Excellency the Commander-in-Chief, 26 April 1863. MSA Military Department (1863–1867), vol. 1133.
122. Act XXII of 1864. Passed by the Governor-General of India in Council. MSA Military Department (1864–1891), vol. 1229.
123. Ibid.
124. In this context, 'contagious diseases' meant venereal disease.
125. Letter to Stuart Hogg, Commissioner of Police from C. Fabre-Tonnerre, 16 September 1867. Proceedings, nos. 112–115(A), 20 February 1869. NAI Home (Public). My italics.
126. Ibid.
127. Banerjee, *Dangerous Outcast*, 153.
128. Banerjee, *Dangerous Outcast*, 162.

Conclusion

1. Indeed, this is even more ironic given that in creating the system, the officers explicitly sought an alternative to the unpleasant treatments for the men.
2. Peers, *Between Mars and Mammon*, 2.

Appendix 1: [Extract] Queries respecting the human race, physical characters

1. 'Queries Respecting the Human Race, to be addressed to Travellers and others. Drawn up by a Committee of the British Association for the Advancement of Science, appointed in 1839, and circulated by the Ethnographical Society of London', 921–23.

Glossary

Abkari	Range of intoxicating substances over which the Company maintained a monopoly
Arrack	A liquor usually made from an extract of sugarcane, rice or palm trees
Ayurveda	Hindu medical practices
Batta	An extra allowance made to soldiers and officers for service in the field
Begum	Lady of rank; mistress
Bhadralok	'Respectable-folk'; Bengali middle-class
Bibi	Lady; mistress
Cantonment	Semi-permanent military station
Chaukidar/Chowkidar	Watchman
Chowdranee/Chaudhryan	Matron
Chowdry	Headman
Chupprassy	Peon or Messenger
Daktar/Daktari	Indian medical men trained in 'western' medicine; usually at one of the medical colleges or schools
Devadasi	'Hand-maiden of the gods'; umbrella term for temple dancers
Dhobi	Washerman
Dhooly/Doolie	A stretcher used to transport the sick or injured
Dram	Unit of volume; approximately 30ml
Eurasian	Person of European and Indian parentage; often referred to in contemporary literature as 'half-cast'
Griffin	A new arrival to India; those unaccustomed to India
Hakim	Physician; doctor of unani-tibb
Kacheri	Public office; place for doing business
Kazi	Public notary; Islamic judge
Kotwal	Police chief; superintendent of police
Inam	Grant of land which was partially or fully free from taxes; often religiously associated
Mofussil	Countryside
Munshi	A writer
Peon	Orderly; man in police employ
Sepoy	Indian soldier in the employ of European forces
Siddha	Tamil medical practices
Tawaif	Highly cultured courtesan
Toddy	A liquor made from the fermented sap of a palm tree
Thana	Administrative area; police post
Unani-tibb	Islamic medical practices
Vaidyas/Vaids	Physician; doctor of ayurvedic practice

Bibliography

I Primary sources

A Manuscripts

Britain

Asia Pacific and Africa Collections, British Library, London

Government of India Papers

Bengal Dispatches
Bengal Judicial
Bengal Military
Bengal Proceedings and Consultations
Bengal Wills
Board of Commissioners for the Affairs of India Collections (also known as Board's Collections)
Bombay Dispatches
Bombay Judicial
Bombay Military
Bombay Proceedings and Consultations
Bombay Public
General Orders
Haileybury Papers
Half-Yearly Reports of the Government Charitable Dispensaries
India Dispatches
Lock Hospital Reports
Madras Dispatches
Madras Judicial
Madras Military
Madras Proceedings and Consultations
Madras Public
Parliamentary Papers

European Manuscripts-Private Papers (MSS Eur)

Compton Papers
Egglestone Papers
Hickson Papers
Porter Papers

Council for World Mission Archives, London

North India, Bengal. Incoming Correspondence
North India, Journals.

National Army Museum Archives, London

Letters and Correspondence

India

Maharashtra State Archives, Mumbai

Military Department
Publications

National Archives of India, New Delhi

Foreign (Secret)
Home (Education)
Home (Judicial)
Home (Medical)
Home (Military)
Home (Public)
Home (Sanitary)
Military (Miscellaneous)

Tamil Nadu State Archives, Chennai

Government Proceedings
Military Consultations
Public Consultations
Surgeon General's Records

Victoria Memorial Library, Kolkata

Hyde Papers

West Bengal State Archives, Kolkata

General Proceedings
General (Sanitation) Proceedings
Judicial Proceedings
Municipal (Sanitation) Proceedings

B Government acts, reports and regulations

Act XLV of 1860, Indian Penal Code.

Annual Report on the Military Lock Hospitals of the Madras Presidency for the Year 1871 (Madras, 1871).

Memorandum on Measures Adopted for Sanitary Improvements in India from June 1869 to June 1870; Together with Abstracts of the Sanitary Reports for 1868 Forwarded from Bengal, Madras and Bombay (London, 1870).

Memorandum on Measures Adopted for Sanitary Improvements in India up to the End of 1867; Together with Abstracts of the Sanitary Reports Hitherto Forwarded from Bengal, Madras and Bombay (London, 1868).

Report of the Commissioners Appointed to Inquire into the Sanitary State of the Army in India; with Précis of Evidence (London, 1863).

Report of the Commissioners Appointed to Inquire into the Organization of the Indian Army (London, 1859).

Report of a Committee of the Associate Medical Members of the Sanitary Commission on the Subject of Venereal Diseases, with Special Reference to Practice in the Army and Navy (1862).

Report on Measures Adopted for Sanitary Improvements in India, from June 1872 to June 1873. Together with Abstracts of Sanitary Reports for 1871, Forwarded from Bengal, Madras, and Bombay (London, 1873).

Rules Establishing Regimental Savings Banks in the Regiments of the British and Indian Armies Serving in Bengal, with the Forms in Use (Calcutta, 1860).

The Regulations of the Government of Fort William in Bengal, in Force at the End of 1853; to Which Are Added, the Acts of the Government of India in Force in That Presidency (London, 1854).

C Contemporary journals

Asiatick Researches
Bombay Quarterly Review
British Medical Journal
Calcutta Journal of Natural History
Calcutta Medical and Physical Society Quarterly Journal
Calcutta Monthly Journal
Calcutta Review
Friend of India
Gleanings in Science
Journal of the Asiatic Society
Journal of the Statistical Society of London
The Indian Annals of Medical Science
The Indian Journal of Medical and Physical Science
The Indian Medical Journal
The Madras Quarterly Medical Journal
The Pamphleteer
Transactions of the Medical and Physical Society of Bombay
Transactions of the Medical and Physical Society of Calcutta

D Published primary sources

Abernethy, John, *Surgical Observations on Diseases Resembling Syphilis: and on Diseases of the Urethra* (London, 1810).

Acton, William, 'Observations on Venereal Diseases in the United Kingdom,' *The Lancet* II (1846), 369–372.

——, *Prostitution: Considered in its Moral, Social & Sanitary Aspects in London and other Large Cities with Proposals for the Control and Prevention of its Attendant Evils* (London, 1857).

Amos, Sheldon, *A Comparative Survey of Laws in Force for the Prohibition, Regulation, and Licensing of Vice in England and Other Countries* (London, 1877).

Andrew, Elizabeth and Bushnell, Katharine, *Facts Recorded by Eye-Witnesses In Regard to the Military Regulation of Vice in India: Being Speeches* (London, 1893).

——, *The Queen's Daughters in India* (London, 1899).

Annesley, James, *Sketches of the Most Prevalent Diseases of India, Comprising a Treatise on Epidemic Cholera of the East* (London, 1825).

Arnott, F. S., 'Report on the Health of the 1st Bombay European Regiment (Fusiliers), from 1st April 1846 to 31st March 1854,' *Transactions of the Medical and Physical Society of Bombay* II (1855), 102–153.

Atkinson, George Francklin, *'Curry and Rice', on Forty Plates: Or, the Ingredients of Social Life at 'Our Station' in India* (London, 1859).

Badenach, Walter, *Inquiry into the State of the Indian Army, with Suggestions for its Improvement, and the Establishment of a Military Police for India* (London, 1826).

Baird, J.G., ed. *Private Letters of the Marquess of Dalhousie* (London, 1910).

Balfour, Andrew, *Health Problems of the Empire: Past Present and Future* (London, 1924).

Ballingall, George, *A Probationary Essay on Syphilis: Submitted to the Royal College of Surgeons of Edinburgh* (Edinburgh, 1820).

——, *Outlines of the Course of Lectures on Military Surgery, Delivered in the University of Edinburgh* (Edinburgh, 1833).

Beames, John, *Memoirs of a Bengal Civilian: The Lively Narrative of a Victorian District Officer* (London, 1984).

Bellew, Captain, *Memoirs of a Griffin; or, A Cadet's First Year in India* (London, 1843).

Belnos, S.C., *Twenty-four Plates Illustrative of Hindoo and European Manners in Bengal* (London, 1832).

Blumenbach, Johann Friedrich, 'On the Natural Varieties of Mankind,' in *The Anthropological Treatises of Johann Friedrich Blumenbach*, ed. & trans. Thomas Bendyshe (Boston, 1865).

Bose, Shib Chander, *The Hindoos As They Are* (London, 1883).

Buchanan, Francis, *A Journey from Madras Through the Countries of Mysore, Canara, and Malabar* (London, 1807).

Buffon, Georges Louis Leclerc, 'The Natural History of the Horse, to which is added, that of the Ass, Bull, Cow, Ox, Sheep, Goat, and Swine' in *Race: The Origins of an Idea, 1760–1850*, ed. Hannah Augstein (Bristol, 2000), 1–9.

Butler, Josephine, *Personal Reminiscences of a Great Crusade* (London, 1898).

Carey, W.H., *The Good Old Days of Honourable John Company; Being Curious Reminiscences Illustrating Manners and Customs of the British in India During the Rule of the East India Company, From 1600 to 1858* (Simla, 1882).

'Census of the Population of the City and District of Murshedabad, Taken in 1829,' *Journal of the Asiatic Society* 2 (1833), 567–569.

Chamber's Encyclopaedia, a Dictionary of Universal Knowledge for the People Illustrated, Vol V. (Edinburgh, 1863).

Chevers, N., 'On the Means of Preserving the Health of European Soldiers in India,' *Indian Annals of Medical Science* 5 (1858) 5, X, 632–762; 6, XI, 203–370; 6, XII, 577–812.

——, *An Inquiry into the Health of Sailors, with Special Reference to the Port of Calcutta* (Calcutta, 1864).

——, *A Manual of Medical Jurisprudence in India* (Calcutta, 1870).

Clark, John, 'Report on Syphilis in HM 13th Light Dragoons,' *The Madras Quarterly Medical Journal* 1 (1839), 370–410.

'Clinical Notes – the Contagious Diseases Act,' *The Indian Medical Journal* 8 (1888), 353–354.

Combe, William (Quiz), *The Grand Master; or, Adventures of Qui Hi? in Hindostan. A Hudibrastic Poem in Eight Cantos* (London, 1816).

Corbyn, Frederick, 'Medical College – Defects in its Constitution,' *The India Journal of Medical and Physical Science* 1,4 (1836).

Cowper, G.A., 'Identity of Syphilis and Gonorrhoea,' *Calcutta Medical and Physical Society Quarterly Journal,* Vol I. (1837), 207–208.

Crook, W., 'Entry on Indian Prostitution,' in *Encyclopaedia of Religion and Ethics*, ed. James Hastings (Edinburgh, 1918), 406–408.

Darwin, Charles, *On the Origin of Species by Means of Natural Selection or, The Preservation of Favoured Races in the Struggle for Life* (London, 1859, reprint 2004).

Doss, Cally Coomar, *The Pamphleteer* 1 (1850).

Dubois, J. A., *Description of the Character, Manners and Customs of the People of India: And of Their Institutions, Religious and Civil* (London, 1817).

Dudley, Sheldon, 'The Prevention of Venereal Disease in the Royal Navy,' Paper presented at the *Inter-Allied Conference on War Medicine, Royal Society of Medicine* (1944).

Edwardes, S. M., *Crime in India: A Brief Review of the More Important Offences Included in the Annual Criminal Returns with Chapters on Prostitution & Miscellaneous Matters* (London, 1924).

Eliot, John, 'Observations on the Inhabitants of the Garrow Hills, Made During a Publick Deputation in the Years 1788 and 1789,' *Asiatic Researches* 3 (1792), 17–37.

'The European Soldier in India,' *Calcutta Review* 59 (1858), 121–148.

Falconer, W., *Remarks on the Influence of Climate, Situation, Nature of Country, Population, Nature of Food, and Way of Life, on the Disposition and Temper, Manners and Behaviour, Intellects, Laws and Customs, Forms of Government, and Religion, of Mankind* (London, 1781).

Fay, Eliza, *Original Letters from India; Containing a Narrative of a Journey through Egypt, and the Author's Imprisonment at Calicut by Hyder Ally* (Calcutta, 1817).

Fayrer, Joseph, 'On the Climate and Fevers of India,' in *A System of Medicine*, ed. Thomas Allbutt (London, 1897), 295–352.

Forbes, James, *Oriental Memories: Selected and Abridged from a Series of Letters Written during Seventeen Years Residence in India including Observations on Parts of Africa and South America, and a Narrative of Occurrences in Four India Voyages* (London, 1813).

Forster, G., *Sketches of the Mythology and Customs of the Hindoos* (London, 1785).

Fox, J., *Britain's Army in India: From its Origins to the Conquest of Bengal* (Allahabad, 1908).
Fuller, Jenny, *The Wrongs of Indian Womanhood* (New York, 1900).
Geddes, William, *Clinical Illustrations of the Diseases of India: As Exhibited in the Medical History of a Body of European Soldiers for a Series of Years From their Arrival in that Country* (London, 1846).
Ghosh, J.N., *Social Evil in Calcutta and Method of Treatment* (Calcutta, 1923).
Grose, John Henry, *A Voyage to the East-Indies; Began in 1750: With Observations Continued till 1764; Including Authentic Accounts of the Mogul Government in General...of the European Settlements, etc.* (London, 1766).
Halhed, Nathaniel, ed., *A Code of Gentoo Laws, or, Ordinations of the Pundits, from a Persian Translation, Made from the Original, Written in the Shanscrit Language* (London, 1776).
Hallet, Job, *De Venereo Morbo; or a Practical Treatise on the Venereal Disease, under all its Various Stages; Containing Instructions to the Afflicted, Rendered so Plain and Easy that they may Assist Themselves in Performing a Cure* (Manchester, 1819).
Hassan Ali, Meer, *Observations of the Mussulmauns of India: Descriptive of their Manners, Customs, Habits and Religious Opinions. Made During a Twelve Years' Residence in their Immediate Society* (London, 1832).
Havelock, Henry, *Memoirs of Major General Sir Henry Havelock,* ed. John Clark Marshman. (London: Longman, Green, Longman and Roberts, 1860).
Heber, Reginald, *A Narrative of a Journey through the Upper Provinces of India, from Calcutta to Bombay, 1824–25* (London, 1828).
Helfer, John William, 'Third Report on Tenasserim – the Surrounding Nations, Inhabitants, Natives and Foreigners, Character, Morals and Religion,' *Journal of the Asiatic Society* VIII (1839), 973–1005.
Hodges, William, *Travels in India during the years 1780, 1781, 1782 & 1783* (London, 1793).
Hodgson, Brian Houghton, 'Tibetan Type of Mankind,' *Journal of the Asiatic Society* 17 (1848), 222–223.
Hodgson, Brian Houghton, 'A Brief Note on Indian Ethnology,' *Journal of the Asiatic Society* 18 (1849), 238–246.
——, 'On the Kocch, Bodo, and Dhimal Tribes,' in *Miscellaneous Essays Relating to Indian Subjects* (London, 1880), 1–160.
Hough, William, *The Practice of Courts-Martial, also the Legal Exposition and Military Explanation of the Military Act, and Articles of War, Together with the Crimes and Sentences of Numerous Courts-Martial, and the Remarks Thereupon by His Majesty, and the Several Commanders in-Chief in the East Indies, and on Foreign Stations, &c., the Whole Forming a Manual of the Judicial and Military Duties of an Officer in Various Situations* (London, 1825).
Hunter, John F. R. S., and Adams, Joseph M. D. F. L. S., *A Treatise on the Venereal Disease; With an Introduction and Commentary by Joseph Adams* (London, 1818).
Hutton, Captain Thomas, 'On the Creation, Diffusion, and Extinction, of Organic Beings,' *The Calcutta Journal of Natural History* 1 (1841), 461–488.
——, 'The Chronology of Creation; or, Geology and Scripture Reconciled,' *Calcutta Review* 14 (1850), 221–264.
Johnson, James, *The Influence of Tropical Climates, More Especially the Climate of India, on European Constitutions* (London, 1813).

Jones, William, 'The Third Anniversary Discourse, Delivered 2 February 1786,' *Asiatic Researches* 1 (1788), 415–431.
——, 'Discourse the Ninth. On the Origin and Families of Nations. Delivered 23 February 1792,' *Asiatic Researches* 2 (1792), 479–492.
Kipling, Rudyard, *Barrack-Room Ballads and Other Verses* (London, 1892).
Lanereaux, E. Trans. G. Whitley, *A Treatise on Syphilis* (London, 1869).
'Lock Hospital Examinations,' *The Indian Medical Journal* 8 (1888), 118–119.
Logan, William, *The Great Social Evil: Its Causes, Extent, Results, and Remedies* (London, 1871).
Long, Rev. James, *Selections from Unpublished Records of Government for the Years 1748 to 1767, Inclusive, Relating Mainly to the Social Condition of Bengal; with a Map of Calcutta in 1784* (Calcutta, 1869).
Lowell, Abbott Lawrence, *Colonial Civil Service. The Selection and Training of Colonial Officials in England, Holland, and France...With an account of the East India College at Haileybury, 1806–1857, by H. M. Stephens* (New York, 1900).
Lushington, Charles, *The History, Design and Present State of the Religious, Benevolent and Charitable Institutions, Founded by the British in Calcutta and its Vicinity* (Calcutta, 1824).
Mackenzie, John, *Army Health in India: Hygiene and Pathology, etc.* (London, 1929).
Mackinnon, Kenneth, *A Treatise on the Public Health, Hygiene and Prevailing Diseases of Bengal and the North-West Provinces* (Cawnpore, 1848).
Macmullen, John Mercier, *Camp and Barrack Room; or, the British Army as it is, by a Late Staff Sergeant of the 13th Light Infantry* (London, 1846).
MacMunn, George Sir, and Lovett, A. C., *The Armies of India* (Bristol, 1911, reprint 1984).
Macpherson, Hugh, 'Analysis of the Later Medical Returns of European Troops Serving in the Bengal Presidency,' *Indian Annals of Medical Science* 4 (1857), 575–607.
Mani, M. S., *The Pen Pictures of the Dancing Girl. With a Sidelight on the Legal Profession* (Madras, 1926).
Marshall, Lieutenant-Colonel James, 'Prevention of Venereal Disease in the British Army,' Paper presented at the *Inter-Allied Conference on War Medicine, Royal Society of Medicine* (1944).
Martin, James Ranald, *Notes on the Medical Topography of Calcutta* (Calcutta, 1837).
——, *The Influence of Tropical Climates on European Constitutions, Including Practical Observations on the Nature and Treatment of the Diseases of Europeans on their Return from Tropical Climates* (London, 1856).
'Medical Jurisprudence-Phrenology,' *The India Journal of Medical and Physical Science* 1, 1 (1834), 18–19.
'Meeting of the Medical and Physical Society,' *Gleanings in Science* 1 (1829), 342.
Mill, James, *The History of British India* (London, 1818).
Mill, John Stuart, *On Liberty* (London, 1859).
Miller, James, *Prostitution Considered in Relation to its Cause and Cure* (Edinburgh, 1859).
Miller, J. R., 'Annual Report of the 23rd Regt. N.L.I.,' *Medical and Physical Society of Bombay* 4 (1858), 275.
Monier-Williams, Sir Monier, *A Few Remarks on the Use of Spirituous Liquors Among the European Soldiers, and on the Punishment of Flogging in the Native Army of the Honourable the East India Company* (London, 1823).

Mundy, Captain, *Pen and Pencil Sketches, Being the Journal of a Tour in India* (London, 1832).
Murdoch, J. C., *Nautch Women: An Appeal to English Ladies on Behalf of their Indian Sisters* (Madras, 1893).
Murray, John, 'Observations Continued throughout 6 years, Relative to the Sick of H.M. 45th Regt. Of Foot: viz. from the 1st January 1832, to the 14th of November 1837; comprising a brief Medical History of the Regiment,' *The Madras Quarterly Medical Journal* 1, 2 (1839), 102–150.
——, *On the Topography of Meerutt, and the Principal Diseases which Prevailed in the 1st Brigade of Horse Artillery at that Place* (Calcutta, 1839).
——, 'Quarterly Reports on the Health of Her Majesty's Troops in the Madras Command for 1838, by the Deputy Inspector General of Hospitals,' *The Madras Quarterly Medical Journal* 1, 2 (1839), 435–443.
'Obituary. Norman Chevers, C.I.E., M.D., F.R.C.S. Eng, Deputy Surgeon General, H.M. Indian Army,' *The British Medical Journal*, 2, 1355 (1886), 1245.
'On the Character of the People, Falsehood, Forgery, Perjury,' *The Calcutta Monthly Journal* 3, Vol I., (1835), 45–60.
'On the Medical Topography of Calcutta,' *Calcutta Medical and Physical Society Quarterly Journal* 4 (1837), 646–658.
Opinions on the Nautch Question Collected and Published by the Punjab Purity Association (Lahore, 1894).
Parkes, Fanny, *Wanderings of a Pilgrim, in Search of the Picturesque, During Four-and-Twenty years in the East; with Revelations of Life in the Zenana* (London, 1850).
Pearman, John, and Paget, George Charles Henry Victor, Marquess of Anglesey, *Sergeant Pearman's Memoirs: Being, Chiefly, His Account of Service with the Third (King's Own) Light Dragoons in India, from 1845 to 1853, Including the First and Second Sikh Wars* (London, 1968).
'Phrenological Development of the Bengalees,' *The Pamphleteer* 1 (1850).
Piddington, Henry, *A Letter to the European Soldiers in India, on the Substitution of Coffee for Spirituous Liquors* (Calcutta, 1839).
Prichard, James Cowles M. D. F. R. S., *Researches into the Physical History of Man.* ed. George W. Stocking, Jr. (Chicago, reprint 1973).
'Queries Respecting the Human Race, to Be Addressed to Travellers and Others. Drawn up by a Committee of the British Association for the Advancement of Science, Appointed in 1839, and Circulated by the Ethnographical Society of London,' *Journal of the Asiatic Society* 13 (1844), 919–932.
Remarks on the Exclusion of Officers of His Majesty's Service from the Staff of the Indian Army' and on the Present State of the European Soldier in India, Whether as Regards His Services, Health, or Moral Character; with a Few of the Most Eligible Means of Modifying the One and Improving the Other, Advocated and Considered (London, 1825).
'Report on the Sickness and Mortality of Troops in the Madras Presidency,' *Journal of the Statistical Society London* 3, 2 (1840), 113–142.
Richards, Frank, *Old Soldier Sahib* (London, 1966).
Roberts, Emma, *Scenes and Characteristics of Hindostan, with Sketches of Anglo-Indian Society* (London, 1835).
Ross, Surgeon-Major, *A Sketch of the Medical History of the Native Army of Bengal for 1868* (1869).

'Rules and Regulations of the Honourable East India Company's Seminary at Addiscombe, 1834,' *Calcutta Review* 2 (1844), 121–152.
Russell, William Howard, *My Indian Mutiny Diary*, ed. Michael Edwardes. (London, 1860, reprint 1957).
Ryan, Michael, *Prostitution in London, with a Comparative View of that in Paris and New York* (London, 1839).
Sanger, William, *The History of Prostitution: Its Extent, Causes, and Effects Throughout the World* (New York, 1910).
Shanks, Dr Archibald, 'Report of H.M. 55th Foot,' *The Madras Quarterly Medical Journal* 1, 2 (1839), 237–268.
Shortt, John M. D., Surgeon-General Superintendent of Vaccination, Madras Presidency, 'The Bayadere; or, Dancing Girls of Southern India,' *Memoirs Read Before the Anthropological Society of London 1867–69*, 3 (1870), 182–194.
Skey, Frederic Carpenter, *A Practical Treatise on the Venereal Disease. Founded on Six Lectures on that Subject, Delivered in the Session of 1838–39, at the Aldersgate School of Medicine. With Plates* (London, 1840).
'Soldiers; Their Morality and Mortality,' *Bombay Quarterly Review* 3 (1855), 167–218.
Statham, J., *Indian Recollections* Vol 3 (London, 1832).
Steel, Flora Annie, *India Through the Ages: A Popular and Picturesque History of Hindustan* (London, 1908).
Stephens, H.M., 'The East India College at Haileybury,' in *Colonial Civil Service. The Selection and Training of Colonial Officials in England, Holland, and France... With an Account of the East India College at Haileybury, 1806–1857*, ed. Abbott Lawrence Lowell (New York, 1900), 233–346.
Thornton, Robert John, *The Medical Guardian of Youth; or, a Popular Treatise on the Prevention and Cure of the Venereal Disease, so that Every Patient May Act for Himself: Containing the Most Approved Practice, and the Latest Discoveries in this Branch of Medicine* (London, 1816).
Travers, Benjamin, *Observations on the Pathology of Venereal Affections* (London, 1830).
Trevelyan, Sir Charles E., K.C.B., *The British Army in 1868* (London, 1868).
Turner, Daniel, *Syphilis. A Practical Dissertation on the Venereal Disease. In which, After an Account of its Nature and Original, the Diagnostick and Prognostic Signs, with the Best Ways of Curing that Distemper* (London, 1724).
A Visit to Madras; Being a Sketch of the Local and Characteristic Peculiarities of That Presidency in the Year 1811 (London, 1821).
'Wallace's Lectures on Syphilitic Eruptions,' *Calcutta Medical and Physical Society Quarterly Journal* 2 (1837), 197–201.
Ward, William, *Account of the Writings, Religion, and Manners, of the Hindoos: Including Translations from their Principal Works* (Serampore, 1811).
Weitbrect, Mrs, *The Women of India, and Christian Work in the Zenana* (London, 1875).
Wellesley, Arthur, Duke of Wellington, 'Memorandum on the Proposed Plan for Altering the Discipline of the Army, 22 April 1829.' *Dispatches, Correspondence, and Memoranda*, Vol 5 (London, 1867).
Williamson, Captain Thomas, *The East India Vade-Mecum; or, Complete Guide to Gentlemen Intended for the Civil, Military, or Naval Service of the Hon. East India Company* (London, 1810).
Yule, Henry; Burnell, Arthur, and Crooke, W., *Hobson-Jobson: A Glossary of Anglo-Indian Words and Phrases: And of Kindred Terms, Etymological, Historical, Geographical, and Discursive* (London, 1886).

II Secondary sources

A Published sources

Alavi, Seema, *The Sepoys and the Company: Tradition and Transition in Northern India, 1770–1830* (Delhi, 1995).

Amin, Sonia, *The World of Muslim Women in Colonial Bengal, 1876–1939* (Leiden, 1996).

Anantha Krishna Iyer, L. K., 'Devadasis in South India: Their Traditional Origin and Development,' *Man in India* 7 (1927), 47–52.

Anderson, Benedict, *Imagined Communities: Reflections on the Origin and Spread of Nationalism* (London, 1983).

Archer, Mildred, 'The Social History of the Nautch Girl,' in *The Saturday Book*, ed. John Hadfield (Essex, 1962), 243–254.

Arnold, David, 'European Orphans and Vagrants in India in the Nineteenth Century,' *Journal of Imperial and Commonwealth History* VII (1979), 104–127.

——, 'Medical Priorities and Practice in 19th Century British India,' *South Asia Research* 5 (1985), 167–183.

——, *Colonizing the Body: State Medicine and Epidemic Disease in Nineteenth Century India* (Berkeley, 1993).

——, 'The Colonial Prison: Power, Knowledge and Penology in Nineteenth-Century India,' in *Subaltern Studies VIII: Essays in Honour of Ranajit Guha*, eds. David Arnold, David Hardiman and Ranajit Guha (Delhi, 1994), 148–157.

——, 'Hunger in the Garden of Plenty: The Bengal Famine of 1770,' ed. Alessa Johns, *Dreadful Visitations: Confronting Natural Catastrophe in the Age of Enlightenment*, (London, 1999), 81–112.

——, *Science, Technology, and Medicine in Colonial India* (Cambridge, 2000).

——, 'Race, Place and Bodily Difference in Early Nineteenth-Century India,' *Historical Research* 77 (2004), 254–273.

Attewell, Guy, *Refiguring Unani Tibb: Plural Healing in Late Colonial India* (Hyderabad, 2007).

Bala, Poonam, *Imperialism and Medicine in Bengal* (New Delhi, 1991).

——, *Medicine and Medical Policies in India: Social and Historical Perspectives* (Plymouth, 2007).

Ballantyne, Tony, *Orientalism and Race: Aryanism in the British Empire* (Hampshire, 2002).

Ballhatchet, Kenneth, *Race, Sex and Class under the Raj: Imperial Attitudes and Policies and their Critics, 1793–1905* (London, 1980).

Banerjee, Sumanta, 'Bogey of the Bawdy-The Changing Concept of "Obscenity" in Nineteenth Century Bengali Culture,' *Economic and Political Weekly* 22, 29 (1987), 1197–1206.

——, 'The 'Beshya' and the 'Babu'-Prostitute and Her Clientele in 19th Century Bengal,' *Economic and Political Weekly* 28, 45 (1993), 2461–2471.

——, *Dangerous Outcast: The Prostitute in Nineteenth Century Bengal (Also Titled: Under the Raj: Prostitution in Colonial Bengal)* (New York, 1998).

Banton, Michael, *Racial Theories* (Cambridge, 1998).

Basalla, George, 'The Spread of Western Science,' *Science* 156 (1967), 611–622.

Bayly, C.A., *Indian Society and the Making of the British Empire: The New Cambridge History of India.* (Cambridge, 1988).

——, 'The British Military-Fiscal State and Indigenous Resistance: India 1750–1820,' in *An Imperial State at War: Britain from 1689 to 1815*, ed. Lawrence Stone (London, 1993), 322–354.
——, 'Knowing the Country: Empire and Information in India,' *Modern Asian Studies* 27 (1993), 3–43.
——, *Origins of Nationality in South Asia: Patriotism and Ethical Government in the Making of Modern India* (Oxford, 1998).
——, *Rulers, Townsmen and Bazaars: North Indian Society in the Age of British Expansion 1770–1870* (New Delhi, 2002).
Bayly, Susan, 'Caste and "Race" in the Colonial Ethnography of India,' in *The Concept of Race in South Asia*, ed. Peter Robb (Oxford, 1996), 165–219.
Bearce, George, *British Attitudes toward India 1784–1858* (Oxford, 1961).
Berger, Mark, 'Imperialism and Sexual Exploitation: A Response to Ronald Hyam's "Empire and Sexual Opportunity",' *Journal of Imperial and Commonwealth History* 17 (1988), 83–89.
Bhattacharya, Rimli, 'The Nautee in "the Second City of the Empire"' *Indian Economic and Social History Review* 40, 2 (2003): 191–235.
Blanco, Richard L., 'Attempted Control of Venereal Disease in the Army of Mid-Victorian England,' *Journal of the Society for Army Historical Research* 45, 184 (1967), 234–241.
Bourne, J.M., 'The East India Company's Military Seminary, Addiscombe, 1809–1858,' *Journal of the Society for Army Historical Research* 57 (1979), 206–222.
Brantlinger, Patrick, *Rule of Darkness: British Literature and Imperialism, 1830–1914* (London, 1988).
Bravo, Michael, 'Ethnological Encounters,' in *Cultures of Natural History*, eds. James A. Secord, E. C. Spary and Nicholas Jardine (Cambridge, 1996), 338–357.
Brimnes, Niels, 'Variolation, Vaccination and Popular Resistance in Early Colonial South India,' *Medical History* 48 (2004), 199–228.
Bristow, Edward, *Vice and Vigilance: Purity Movements in Britain since 1700* (Dublin, 1978).
'British Siblings in Bihar in Search of their Roots,' *The Bihar Times*, 12 January 2009.
Bubb, Alexander, 'The Life of the Irish Soldier in India: Representations and Self-Representations, 1857–1922' *Modern Asian Studies* 46, 4 (2012), 769–813.
Burton, Antoinette, 'The White Woman's Burden: British Feminists and the "Indian Woman" 1865–1915,' in *Western Women and Imperialism*, eds. Nupur Chaudhury and Margaret Strobel (Bloomington, 1992), 137–157.
——, *Burdens of History: British Feminists, Indian Women, and Imperial Culture, 1865–1915* (Chapel Hill; London, 1994).
——, (ed.) *Politics and Empire in Victorian Britain, A Reader* (New York, 2001).
Callahan, Raymond, *The East India Company and Army Reform, 1783–1798* (Cambridge [Mass], 1972).
Cannadine, David, *Ornamentalism: How the British Saw their Empire* (London, 2001).
Carlyle, E.I., 'Tulloch, Sir Alexander Murray (1803–1864),' in *Oxford Dictionary of National Biography*, ed. John Sweetman (London, 2004).
Carroll, Lucy, 'Law, Custom and Statutory Social Reform: "The Hindu Widow Remarriage Act of 1856",' *Indian Economic and Social History Review* 20, 4 (1983), 363–388.

Cassels, Nancy 'Bentinck: Humanitarian and Imperialist-The Abolition of Suttee,' *The Journal of British Studies* 5 (1965), 77–87.
Chakrabarti, Pratik, *Materials and Medicine: Trade, Conquest and Therapeutics in the Eighteenth Century* (Manchester, 2010).
——, '"Neither of Meate nor Drinke, but What the Doctor Alloweth": Medicine Amidst War and Commerce in Eighteenth-Century Madras,' *Bulletin of the History of Medicine* 80 (2006), 1–38.
Chakravarti, Uma, 'Whatever Happened to the Vedic Dasi? Orientalism, Nationalism, and a Script for the Past,' in *Recasting Women: Essays in Colonial History*, eds. Kumkum Sangari and Sudesh Vaid (Delhi, 1990), 27–87.
Chandavarkar, Rajnarayan, 'Plague Panic and Epidemic Politics in India, 1896–1914,' in *Epidemics and Ideas: Essays on the Historical Perception of Pestilence*, eds. Terence Ranger and Paul Slack (Cambridge, 1992), 203–240.
——, 'Customs of Governance: Colonialism and Democracy in Twentieth Century India,' *Modern Asian Studies* 41 (2007), 441–470.
Chatterjee, Indrani, 'Colouring Subalternity: Slaves, Concubines and Social Orphans,' in *Subaltern Studies X: Writings on South Asian History and Society*, eds. Gautam Bhadra, Gyan Prakash and Susie J. Tharu (New Delhi, 1999), 49–97.
——, *Gender, Slavery and Law in Colonial India* (Delhi, 2002).
——, *Unfamiliar Relations: Family and History in South Asia* (London, 2004).
Chatterjee, Indrani, and Eaton, Richard Maxwell, eds. *Slavery & South Asian History* (Bloomington, Ind., 2007).
Chatterjee, Partha, *The Nation and its Fragments: Colonial and Postcolonial Histories* (Princeton, N.J., 1993).
Chatterjee, Ratnabali, 'The Queen's Daughters: Prostitutes as a Social Outcast Group in Colonial India,' *Report, Michelsen Institute* (1992).
——, 'Prostitution in Nineteenth Century Bengal: Constructions of Class and Gender,' *Social Scientist* 21 (1993), 159–172.
Chaudhuri, Nupur, 'Clash of Cultures: Gender and Colonialism in South and Southeast Asia,' in *A Companion to Gender History*, eds. Teresa Meade and Merry Wiesner (Oxford, 2004), 430–443.
Chaudhuri, Nupur, and Strobel, Margaret, eds. *Western Women and Imperialism: Complicity and Resistance* (Bloomington, 1992).
Cohn, Bernard, 'Recruitment and Training of British Civil Servants in India, 1600–1860,' in *Asian Bureaucratic Systems Emergent from the British Imperial Tradition*, ed. Ralph Braibanti (Durham, N.C., 1966), 87–140.
Cominos, Peter, 'Late Victorian Sexual Respectability and the Social System,' *International Review of Social History* 8 (1963), 216–250.
Crawford, Dirom Grey, *A History of the Indian Medical Service, 1600–1913.* (London, 1914).
Crowell, Lorenzo M., 'Military Professionalism in a Colonial Context: The Madras Army, circa 1832,' *Modern Asian Studies* 24 (1990), 249–273.
Curtin, Philip D., *Death by Migration: Europe's Encounter with the Tropical World* (Cambridge, 1989).
Dalrymple, William, *White Mughals: Love and Betrayal in Eighteenth-Century India* (London, 2003).
Dang, Kokila, 'Prostitutes, Patrons and the State,' *Social Scientist* 21 (1993), 173–196.

Davidson, Roger, and Hall, Lesley, eds. *Sex, Sin and Suffering: Venereal Disease and European Society since 1870* (London, 2001).
Davin, Anna, 'Imperialism and Motherhood' *History Workshop Journal* 5 (1978): 9–65.
Davis, Mike, *Late Victorian Holocausts: El Nino Famines and the Making of the Third World* (London, 2000).
de Watteville, Herman, *The British Soldier. His Daily Life from Tudor to Modern Times* (London, 1954).
Dewey, Clive, *Anglo-Indian Attitudes: The Mind of the Indian Civil Service* (London, 1993).
Dirks, Nicholas, *Castes of Mind: Colonialism and the Making of Modern India* (Princeton, 2001).
——, *The Scandal of Empire: India and the Creation of Imperial Britain* (Cambridge, [Mass.], 2006).
Dodson, Michael S., *Orientalism, Empire, and National Culture: India, 1770–1880* (Basingstoke: Palgrave Macmillan, 2007).
Dyson, Ketaki Kushari, *A Various Universe: A Study of the Journals and Memoirs of British Men and Women in the Indian Subcontinent, 1765–1856* (Delhi, 1978).
Edwards, Owain, 'Captain Thomas Williamson of India' *Modern Asian Studies* 14, 4 (1980): 673–682.
Ernst, Waltrud, 'Colonial Psychiatry, Magic and Religion. The Case of Mesmerism in British India,' *History of Psychiatry* 15 (2004), 57–71.
Farrant, John H., 'Grose, John Henry (b. 1732, d. in or after 1774),' in *Oxford Dictionary of National Biography* (London, 2004).
Fisch, Jörg, *Cheap Lives and Dear Limbs: The British Transformation of the Bengal Criminal Law, 1769–1817* (Wiesbaden, 1983).
Foucault, Michel, *Discipline and Punish: The Birth of the Prison* (New York, 1979).
——, *History of Sexuality: The Will to Knowledge* (London, 1998).
Ghosh, Durba, 'Making and Un-making Loyal Subjects: Pensioning Widows and Educating Orphans in Early Colonial India,' *Journal of Imperial and Commonwealth History* 31, 1 (2003), 1–28.
——, *Sex and the Family in Colonial India: The Making of Empire* (Cambridge, 2006).
Ghosh, Suresh Chandra, *The Social Condition of the British Community in Bengal 1757–1800* (Leiden, 1970).
——, *Birth of a New India: Fresh Light on the Contributions made by Bentinck, Dalhousie and Curzon in the Nineteenth Century* (Delhi, 2001).
Gilbert, Marc Jason, 'Empire and Excise: Drugs and Drink Revenue and the Fate of States in South Asia,' in *Drugs and Empires: Essays in Modern Imperialism and Intoxication, c. 1500–1930*, eds. James Mills and Patricia Barton (Basingstoke, 2007), 116–141.
Gilbert, Pamela, *Mapping the Victorian Social Body* (Albany, 2004).
Gilfoyle, Timothy, 'Prostitutes in the Archives: Problems and Possibilities in Documenting the History of Sexuality,' *American Archivist* 57 (1994), 514–527.
Gilmour, David, *The Ruling Caste: Imperial Lives in the Victorian Raj* (London, 2005).
Green, Nile, *Islam and the Army in Colonial India: Sepoy Religion in the Service of Empire* (Cambridge, 2009).
Gupta, Charu, *Sexuality, Obscenity, Community: Women, Muslims, and the Hindu Public in Colonial India* (New Delhi, 2001).
Gupta, Partha Sarathi and Deshpande, Anirudh, eds. *The British Raj and Its Indian Armed Forces, 1857–1939* (New Delhi, 2002).

Hall, Catherine, *Civilising Subjects: Metropole and Colony in the English Imagination 1830–1867* (Oxford, 2002).
Hall-Matthews, David, *Peasants, Famine and the State in Colonial Western India* (Basingstoke, 2005).
Harrison, Mark, *Public Health in British India: Anglo-Indian Preventive Medicine, 1859–1914* (Cambridge, 1994).
——, '"The Tender Frame of Man": Disease, Climate, and Racial Difference in India and the West Indies, 1760–1860,' *Bulletin of the History of Medicine* 70, 1 (1996), 68–93.
——, *Climates & Constitutions: Health, Race, Environment and British Imperialism in India 1600–1850* (Oxford, 1999).
——, 'Was there an Oriental Renaissance in Medicine? The Evidence of the Nineteenth-Century Medical Press,' in *Negotiating India in the Nineteenth-Century Media*, eds. David Finkelstein and Douglas M. Peers (Basingstoke, 2000), 233–253.
——, 'Medicine and Orientalism: Perspectives on Europe's Encounter with Indian Medical Systems,' in *Health, Medicine and Empire: Perspectives on Colonial India*, eds. Biswamoy Pati and Mark Harrison (Hyderabad, 2001), 37–87.
Hawes, Christopher, *Poor Relations: The Making of a Eurasian Community in British India, 1773–1833* (Richmond, 1996).
Heesterman, J. C., 'Western Expansion, Indian Reaction: Mughal Empire and British Raj,' in *The Inner Conflict of Tradition: Essays in Indian Ritual, Kingship, and Society* (Chicago, 1985), 158–179.
Hervey, H., *Cameos of Indian Crime* (London, 1950).
Hodges, Sarah, '"Looting" the Lock Hospital in Colonial Madras during the Famine Years of the 1870s,' *Social History of Medicine* 18 (2005), 379–398.
Howell, Philip, *Geographies of Regulation: Policing Prostitution in Nineteenth-century Britain and the Empire* (Cambridge, 2009).
Hutchins, F., *The Illusion of Permanence* (Princeton, 1967).
Hyam, Ronald, *Empire and Sexuality: The British Experience* (Manchester, 1990).
Inden, R., 'Orientalist Constructions of India,' *Modern Asian Studies* 20 (1986), 401–446.
Jacob, T., *Cantonments in India: Evolution and Growth* (New Delhi, 1994).
Jacyna, L. S., 'Abernethy, John (1764–1831),' in *Oxford Dictionary of National Biography* (Oxford, 2004).
James, Lawrence, *Raj: The Making and Unmaking of British India* (London, 1997).
Joardar, Biswanath, *Prostitution in 19th and Early 20th Century Calcutta* (Delhi, 1985).
Joseph, Betty, *Reading the East India Company, 1720–1840: Colonial Currencies of Gender* (Chicago, 2004).
Kaminsky, Arnold, 'Morality Legislation and British Troops in late 19th Century India,' *Military Affairs* 43 (1979), 78–93.
Kejariwal, O. P., *The Asiatic Society of Bengal and the Discovery of India's Past, 1784–1838* (Oxford, 1988).
Kolff, Dirk, *Naukar, Rajput, and Sepoy: The Ethnohistory of the Military Labour Market in Hindustan, 1450–1850* (Cambridge, 1990).
Korieh, Chima, 'Dangerous Drinks and the Colonial State: "Illicit" Gin Prohibition and Control in Colonial Nigeria,' in *Drugs and Empires: Essays in Modern Imperialism and Intoxication, c. 1500–c.1930*, eds. James Mills and Patricia Barton (Basingstoke, 2007), 101–115.

Krishnamurty, J., *Women in Colonial India: Essays on Survival, Work and the State* (Delhi, 1989).
Lawrence, Christopher, *Medicine in the Making of Modern Britain, 1700–1920* (London, 1994).
Legg, Stephen, *Spaces of Colonialism: Delhi's Urban Governmentalities* (Oxford, 2007).
——, 'Governing Prostitution in Colonial Delhi: From Cantonment Regulations to International Hygiene (1864–1939)' *Social History* 34, 4 (2009): 447–67.
Levine, Philippa, 'Venereal Disease, Prostitution and the Politics of Empire: The Case of British India,' *Journal of the History of Sexuality* 4, 4 (1994), 579–602.
——, 'Rereading the 1890s: Venereal Disease as "Constitutional Crisis" in Britain and British India,' *Journal of Asian Studies* 55 (1996), 585–612.
——, 'Public Health, Venereal Disease and Colonial Medicine in the Later Nineteenth Century,' in *Sin, Sex and Suffering: Venereal Disease and European Society Since 1870*, eds. Roger Davidson and Lesley Hall (London, 2001), 160–172.
——, *Prostitution, Race, and Politics: Policing Venereal Disease in the British Empire* (New York, 2003).
——, '"A Multitude of Unchaste Women": Prostitution in the British Empire,' *Journal of Women's History* 15, 4 (2004), 159–163.
Liddle, Joanna, and Joshi, Rami, 'Gender and Imperialism in British India,' *South Asia Research* (1985), 147–165.
Macdonald, Donald, *Surgeons Twoe and a Barber. Being Some Account of the Life and Work of the Indian Medical Service, 1600–1947* (London, 1950).
Majeed, Javed, *Ungoverned Imaginings: James Mill's The History of British India and Orientalism* (Oxford, 1992).
Mallampalli, Chandra, 'Meet the Abrahams: Colonial Law and a Mixed Race Family from Bellary, South India, 1810–63' *Modern Asian Studies* 42, 5 (2008): 929–970.
——, *Race, Religion, and Law in Colonial India: Trials of an Interracial Family* (Cambridge, 2011).
Mani, Lata, *Contentious Traditions: The Debate on Sati in Colonial India* (Berkeley, 1998).
Marglin, Frédérique Apffel, *Wives of the God-King: The Rituals of the Devadasis of Puri* (Delhi, 1985).
Marshall, Peter, 'British Society in India under the East India Company,' *Modern Asian Studies* 31, 1 (1997), 89–108.
McHugh, Paul, *Prostitution and Victorian Social Reform* (London, 1980).
Misra, Bankey Bihari, *The Central Administration of the East India Company, 1773–1834* (Manchester, 1959).
Moore, R.J., 'The Abolition of Patronage in the Indian Civil Service and the Closure of Haileybury College,' *The Historical Journal* 7,2 (1964), 246–257.
Mukharji, Projit Bihari, *Nationalizing the Body: The Medical Market, Print and Daktari Medicine* (London, 2009).
Mukherji, S. K., *Prostitution in India* (Calcutta, 1934).
Murphy, Sharon, 'Libraries, Schoolrooms, and Mud Gadowns: Formal Scenes of Reading at East India Company Stations in India, c. 1819–1835', *Journal of the Royal Asiatic Society Series* 3, 21, 4 (2011): 459–467.
Naidis, Mark, 'British Attitudes toward the Anglo-Indians,' *South Atlantic Quarterly* 62 (1963), 407–422.

Nair, Janaki, 'The Devadasi, Dharma and the State' *Economic and Political Weekly* 29, 50 (1994), 3157–3167.
——, *Women and Law in Colonial India: A Social History* (New Delhi, 1996).
Nandy, Ashis, *The Intimate Enemy: Loss and Recovery of Self Under Colonialism* (New Delhi, 1983).
Nandy, Somendra Candra, *Life and Times of Cantoo Baboo (Krisna Kanta Nandy): The Banian of Warren Hastings, 1742–1804* (Bombay, 1978).
Oldenburg, V. T., *The Making of Colonial Lucknow, 1856–1877* (Princeton, 1984).
——, 'Lifestyle as Resistance: The Case of the Courtesans of Lucknow, India,' *Feminist Studies* 16 (1990), 259–288.
Omissi, David, *The Sepoy and the Raj: The Indian Army, 1860–1940* (London, 1994).
Omissi, David, and Killingray, David, *Guardians of Empire: The Armed Forces of the Colonial Powers c. 1700–1964* (Manchester, 1999).
Oriel, J. D., *The Scars of Venus: A History of Venerology* (London, 1994).
Pande, Ishita, *Medicine, Race and Liberalism in British Bengal: Symptoms of Empire* (London, 2010).
Parker, Kunal, '"A Corporation of Superior Prostitutes" Anglo-Indian Legal Conceptions of Temple Dancing Girls, 1800–1914,' *Modern Asian Studies* 32 (1998), 559–633.
Pati, Biswamoy, and Harrison, Mark, eds. *Health Medicine and Empire: Perspectives on Colonial India* (Hyderabad, 2001).
Paxton, Nancy, *Writing under the Raj: Gender, Race, and Rape in the British Colonial Imagination, 1830–1947* (New Brunswick, New Jersey, 1999).
Peers, Douglas, 'The Habitual Nobility of Being: British Officers and the Social Constitution of the Bengal Army in the Nineteenth Century,' *Modern Asian Studies* 25 (1991), 545–569.
——, 'The Thin End of the Wedge: Medical Relativities as a Paradigm of Early Modern Indian-European Relations,' *Modern Asian Studies* 29, 1 (1995), 141–170.
——, *Between Mars and Mammon: Colonial Armies and the Garrison State in India 1819–1835* (London, 1995).
——, 'Sepoys, Soldiers and the Lash: Race, Caste and Army Discipline in India, 1820–1850,' *Journal of Imperial and Commonwealth History* 23 (1995), 211–247.
——, 'Soldiers, Surgeons and the Campaigns to Combat Sexually Transmitted Diseases in Colonial India, 1805–1860,' *Medical History* 42, 2 (1998), 137–160.
——, 'Imperial Vice: Sex, Drink and the Health of the British Troops in North Indian Cantonments, 1800–1858,' in *Guardians of Empire: The Armed Forces of the Colonial Powers c.1700–1964*, eds. David Killingray and David Omissi (Manchester, 1999), 25–52.
——, 'The Raj's Other Great Game: Policing the Sexual Frontiers of the Indian Army in the First Half of the Nineteenth Century,' in *Discipline and the Other Body: Correction, Corporeality, Colonialism*, eds. Steven Pierce and Anupama Rao (Raleigh, 2006), 115–150.
——, 'Gunpowder Empires and the Garrison State: Modernity, Hybridity, and the Political Economy of Colonial India, circa 1750–1860,' *Comparative Studies of South Asia, Africa and the Middle East* 27 (2007), 245–258.
——, '"The more this foul case is stirred, the more offensive it becomes": Imperial Authority, Victorian Sentimentality and the Court Martial of Colonel

Crawley, 1862–4.' *Fringes of Empire: Peoples, Places and Spaces in Colonial India*, eds. Sameeta Agha and Elizabeth Kolsky (Delhi, 2009), 207–235.

Philips, C. H., ed. *The Correspondence of Lord William Cavendish Bentinck, Governor-General of India 1828–1835* (Oxford, 1977).

Potter, David C., *Government in Rural India. An Introduction to Contemporary District Administration* (London, 1964).

Prasad, Awadh Kishore, *Devadasi System in Ancient India: A Study of Temple Dancing Girls of South India* (Delhi, 1990).

Pryor, Alan, 'Indian Pale Ale: An Icon of Empire' *Commodities of Empire, Working Paper Series* 13 (2009).

Punekar, S. D., and Rao, Kamala, *A Study of Prostitutes in Bombay (With reference to Family Background)* (Bombay, 1962).

Raj, Kapil, 'Colonial Encounters and the Forging of New Knowledge and National Identities: Great Britain and India, 1760–1850,' *Osiris* 15 (2000), 119–134.

——, 'The Historical Anatomy of a Contact Zone: Calcutta in the Eighteenth Century' *Indian Economic and Social History Review* 48, 1 (2011), 55–82.

Ramanna, Mridula, *Western Medicine and Public Health in Colonial Bombay* (London, 2002).

Ramasubban, Radhika, 'Imperial Health in British India, 1857–1900,' in *Disease, Medicine, and Empire: Perspectives on Western Medicine and the Experience of European Expansion*, eds. Roy MacLeod and Milton Lewis (London, 1988), 38–60.

Ranger, Terence, and Slack, Paul, eds. *Epidemics and Ideas: Essays on the Historical Perception of Pestilence* (Cambridge, 1992).

Rich, Paul, *Race and Empire in British Politics* (Cambridge, 1990).

Rocher, Rosane, 'British Orientalism in the Eighteenth Century: The Dialectics of Knowledge and Government.' *Orientalism and the Postcolonial Predicament*, eds. Carol Breckenridge and Peter van der Veer (Philadelphia, 1993).

——, 'Sanskrit for Civil Servants 1806–1818,' *Journal of the American Oriental Society* 122, 2 (2002), 381–390.

Rosselli, John, *Lord William Bentinck: The Making of a Liberal Imperialist, 1774–1839* (London, 1974).

——, 'The Self-Image of Effeteness: Physical Education and Nationalism in Nineteenth-Century Bengal,' *Past and Present* 86 (1980), 121–148.

Roy, Kaushik, 'The Armed Expansion of the English East India Company, 1740s–1849,' in *A Military History of India and South Asia: From the East India Company to the Nuclear Era*, eds. Daniel Marston and Chandar Sundaram (London, 2007), 1–15.

Sangari, Kumkum, and Vaid, Sudesh, eds. *Recasting Women: Essays in Indian Colonial History* (Delhi, 1990).

Sarkar, Tanika, *Hindu Wife, Hindu Nation: Community, Religion and Cultural Nationalism* (New Delhi, 2001).

Schofield (nee Butler Brown), Katherine, 'The Social Liminality of Musicians: Case Studies from Mughal India and Beyond' *Twentieth Century Music* 3, 1 (2007), 13–49.

Scott, James C., *Weapons of the Weak: Everyday Forms of Peasant Resistance* (New Haven, 1985).

Seal, Anil, *The Emergence of Indian Nationalism: Competition and Collaboration in the Later Nineteenth Century* (Cambridge, 1971).

Sen, Indrani, *Women and Empire: Representations in the Writings of British India, 1858–1900* (New Delhi, 2002).
Sen, Samita, *Women and Labour in Late Colonial India – The Bengal Jute Industry* (Cambridge, 1999).
Sen, Sudipta, 'Colonial Aversions and Domestic Desires: Blood, Race, Sex and the Decline of Intimacy in Early British India,' in *Sexual Sites, Seminal Attitudes: Sexualities, Masculinities, and Culture in South Asia*, ed. Sanjay Srivastava (New Delhi, 2004), 49–82.
Sharp, Jenny, *Allegories of Empire: The Figure of Woman in the Colonial Text* (Minneapolis, 1993).
Sigsworth, E.M. and Wyke, T.J., 'A Study of Victorian Prostitution and Venereal Disease,' in *Suffer and Be Still*, ed. Martha Vicinus (Bloomington, 1972), 77–99.
Singha, R, ' "Providential" Circumstances: The Thuggee Campaign of the 1830s and Legal Innovation,' *Modern Asian Studies* 27 (1993), 83–146.
——, *A Despotism of Law: British Criminal Justice and Public Authority in North India, 1772–1837* (Oxford, 1998).
Sinha, Mrinalini, *Colonial Masculinity: The 'Manly Englishman' and the 'Effeminate Bengali' in the Late Nineteenth Century* (Manchester, 1995).
——, 'Britishness, Clubbability, and the Colonial Public Sphere: The Genealogy of an Imperial Institution in Colonial India,' *The Journal of British Studies* 40 (2001), 489–521.
Smith, F.B., 'The Contagious Diseases Act Reconsidered,' *Social History of Medicine* 3, 2 (1990), 197–215.
Spear, Percival, *The Nabobs: A Study of Social Life of the English in the Eighteenth Century* (London, 1963).
Spiers, Edward, *The Army and Society 1815–1914* (London, 1980).
Spivak, Gayatri Chakravorty, 'The Rani of Sirmur: An Essay in Reading the Archives,' *History and Theory* 24, 3 (1985), 247–272.
Spongberg, Mary, *Feminizing Venereal Disease: The Body of the Prostitute in Nineteenth-Century Medical Discourse* (London, 1997).
Sreenivas, Mytheli, *Wives, Widows, and Concubines: The Conjugal Family Ideal in Colonial India* (Bloomington, 2008).
Srinivasan, Amrit, 'Reform and Revival: The Devadasi and her Dance,' *Economic and Political Weekly* 20, 44 (1985), 1869–1876.
——, 'Reform or Conformity? Temple "Prostitution" and the Community in the Madras Presidency,' in *Structures of Patriarchy: State, Community and Household in Modernising Asia*, ed. Bina Agarwal (New Delhi, 1991), 175–198.
Stanley, Peter, *White Mutiny: British Military Culture in India, 1825–1875* (London, 1998).
Stein, Burton, *Thomas Munro: The Origins of the Colonial State and his Vision of Empire* (Delhi, 1989).
Stepan, Nancy, *The Idea of Race in Science: Great Britain 1800–1960* (London, 1982).
Stocking, George, *Victorian Anthropology* (New York, 1987).
Stokes, Eric, *The English Utilitarians and India* (Delhi, 1989).
Stoler, Ann Laura, 'Making Empire Respectable: The Politics of Race and Sexual Morality in 20th-Century Colonial Cultures,' *American Ethnologist* 16 (1989), 634–660.
——, 'Rethinking Colonial Categories: European Communities and the Boundaries of Rule,' *Comparative Studies in Society and History* 31, 1 (1989), 135–201.

——, *Race and the Education of Desire: Foucault's 'History of Sexuality' and the Colonial Order of Things* (Durham, 1996).

——, *Carnal Knowledge and Imperial Power: Race and the Intimate in Colonial Rule* (Berkeley, 2002).

Strachan, Hew, *Wellington's Legacy: The Reform of the British Army, 1830–54* (Manchester, 1984).

Streets-Salter, Heather, *Martial Races: The Military, Race and Masculinity in British Imperial Culture, 1857–1914* (Manchester, 2004).

Suleri, Sara, *The Rhetoric of British India* (Chicago, 1992).

Sundar, Pushpa, *Patrons and Philistines: Arts and the State in British India, 1773–1947* (Delhi, 1995).

Sundara Raj, *Prostitution in Madras: A Study in Historical Perspective* (Delhi, 1993).

Tambe, Ashwini, *Codes of Misconduct: Regulating Prostitution in Late Colonial Bombay* (Minneapolis, 2009).

Thomas, Patrick John, *The Growth of Federal Finance in India: Being a Survey of India's Public Finances from 1833 to 1939* (Madras, 1939).

Thompson, Andrew, *The Empire Strikes Back? The Impact of Imperialism on Britain from the Mid-Nineteenth Century* (Harlow, 2005).

Tidrick, Kathryn, *Empire and the English Character* (London, 1990).

Tobin, Beth Fowkes, *Picturing Imperial Power: Colonial Subjects in Eighteenth-Century British Painting* (London, 1999).

Trautmann, Thomas R., *Aryans and British India* (Berkeley 1997).

Travers, Robert, *Ideology and Empire in Eighteenth Century India: The British in Bengal* (Cambridge, 2007).

Tripathi, Amales, *Trade and Finance in the Bengal Presidency, 1793–1833* (Calcutta 1979).

Trustram, Myna, 'Distasteful and Derogatory? Examining Victorian Soldiers for Venereal Disease,' in *The Sexual Dynamics of History: Men's Power, Women's Resistance*, ed. London Feminist History Group (London, 1983), 154–164.

van Heyningen, Elizabeth, 'The Social Evil in the Cape Colony 1868–1902: Prostitution and the Contagious Diseases Acts' *Journal of Southern African Studies* 10, 2 (1984), 170–197.

Vijaisri, Priyadarshini, *Recasting the Devadasi: Patterns of Sacred Prostitution in Colonial South India* (New Delhi, 2004).

Wagner, Kim, *Thuggee: Banditry and the British in Early Nineteenth-Century India* (Basingstoke, 2007).

Walkowitz, Judith, *Prostitution and Victorian Society: Women, Class and the State* (Cambridge, 1980).

Ware, Vron, *Beyond the Pale: White Women, Racism and History* (London, 1992).

Washbrook, David, 'Law, State and Agrarian Society in Colonial India,' *Modern Asian Studies* 15 (1981), 649–721.

——, 'Progress and Problems: South Asian Economic and Social History, 1720–1860', *Modern Asian Studies* 21 (1988), 57–96.

Weeks, Jeffrey, *Sex, Politics and Society: The Regulation of Sexuality since 1800* (London, 1981).

Whitehead, Judy, 'Bodies Clean and Unclean: Prostitution, Sanitary Legislation, and Respectable Femininity in Colonial North India,' *Gender and History* 7, 1 (1995), 41–63.

Zimmerman, Francis, *The Jungle and the Aroma of Meats: An Ecological Theme in Hindu Medicine* (Berkeley, 1988).

B Unpublished dissertations

Chakrabarti, A., 'Widowhood in Colonial Bengal, 1850–1930,' (unpublished PhD thesis, University of Calcutta, 2004).
Kolksky, E., 'The Body Evidencing the Crime: Gender, Law and Medicine in Colonial India,' (unpublished PhD thesis, Columbia University 2002).
Tambe, A., 'Codes of Misconduct: The Regulation of Prostitution in Colonial Bombay, 1860–1947,' (unpublished PhD thesis, American University, 2000).

Index

Act XIV of 1868. *See* contagious diseases; Indian Contagious Diseases Act; Indian Contagious Diseases Act (Act XIV of 1868)
Act XXII of 1864. *See* Cantonment Act (Act XXII of 1864); contagious diseases
Addiscombe (Company Military Seminary), 85
  compared with the Military Academy at Woolwich, 87
  officer training at, 86–87, 88–89
  See also colleges
Ainslie, Whitelaw (surgeon on the Madras Medical Board), 127, 128, 139
alcohol. *See* intemperance; liquor
Anderson, Benedict, 98
Anglicists, 84, 220n1
  See also Bentinck, Lord William; liberalism
Arnold, David, 10, 96
Arnot, F.S., 'Report on the Health of the 1st Bombay Regiment', 137–138
Asiatic Society of Bengal, 98
Ayurveda:
  Company support of, 52, 92–93, 94
  and *daktars*, 93–94
  European interest in, 52–53
  and European medical practices
    asserted as superior over, 98–99
    and the teaching of indigenous practices, 52, 92–93
  vaidyas/vaids (practitioner of ayurvedic medicine)
    European dependence on practices of, 52–53
    European distrust of, 94, 104
  See also indigenous medicine
Ballhatchet, Kenneth, 208n34
Bartle Frere, Henry (Governor of Bombay), 183–184
batta (field pay), 20, 139
Beames, John, 87–88, 89
begums, 29
Bellew, Captain, *Memoirs of a Griffin*, 29
Bengal:
  Bhadralok (educated Bengali middle class), 103
  and the Cantonment Act (Act XXII of 1864), 184
  Charitable Dispensaries in, 174
  'Qui Hi' (soldier in army of), 1, 21, 202n2
  See also Calcutta; Fort William; Presidencies
Bengal Medical Service, 90
Bengal Military Board, 76–77, 219n110
Bentinck, Lord William, 92, 93, 162
  abolition of the lock hospital system by, 9, 79, 82–83, 92, 105, 112, 160–162
Blumenbach, Johan Freidrich, 95–96, 223n55
Bombay:
  Bombay Medical and Physical Society, 98, 120
  'Duck' (soldier in army of), 21
  Henry Bartle Frere (Governor of Bombay), 183–184
  John FitzGibbon (Governor of Bombay), 160
  native dispensary in, 171
  See also Presidencies
Bombay Quarterly Review, 121
Burke, William (Inspector-General of Hospitals):
  lock-hospital system abolition recommended by, 82, 92, 160–165, *163t5.1*

rejection of findings of, 112, 161, 167, 178–179
Butler, Josephine, 54

Calcutta:
Calcutta Medical and Physical Society, 98, 99, 103–104, 103–104, 111
Calcutta Phrenological Society, 101
See also Bengal
Calcutta Madrassa, 52, 92–93
Calcutta Sanskrit College, 52, 92–93
Calcutta University, 93
Cantonment Act (Act XIII of 1889), 159
Cantonment Act (Act XXII of 1864), 10, 123, 184–185
See also contagious diseases acts; punishment
cantonments:
kotwal (superintendent of police), 57–58, 61–62, 76–77, 219n101
military and government regulation of, 131–140
charitable institutions:
Lord Clive's Fund, 33
Military Orphan Society, 26, 33
Chevers, Norman:
career of, 228n4
health of soldiery, proposals by, 120, 124, 127–130, 139
Indian Annals of Medical Science, 180
officers associated with intemperance by, 123–124
children:
Eurasian children, 26, 30, 32, 36
See also orphans
cholera, 53, 94, 120, 130, 162, 179–181
chowdranee/chaudhryan. *See* matrons
Clark, Surgeon John, 111–112, 127
class:
class-based definitions of working-class European soldiers, 3–4, 9, 12–13, 17, 22, 83, 122, 124, 139, 148–149
'low-class' females associated with venereal disease transmission, 8–9, 53–54, 184–185
coffee:
coffee rooms/shops for British soldiers, 22–23, 138
Piddington's proposals, 128–129
co-habitation and companionship:
and higher paid soldiers, 23
in India before 1830, 24–25
in pre-colonial India, 23
See also dancing girls; marriage
colleges:
bonding of men associated with, 85, 89
and Company aims and needs, 85, 86–87
economic benefits of, 87
lack of discipline at, 85, 87–89
See also Addiscombe; education; Fort William College; Haileybury; medical colleges
colonialism:
and the categorisation of various groups of women as 'prostitutes', 12, 15, 18–19, 38–40, 46–47, 56–57
and class-based definitions of European soldiers, 2–4, 9, 12–13
lock hospital system as emblematic of colonial control, 160, 171–176, 189, 194–195
and sexuality, 3–7, 114
See also Company Raj; Crown raj; regulations
Company raj:
fiscal demands of
and expanding political power, 2–4, 7–8, 59–60, 84–85, 90–92
and the families of soldiers, 22–24, 30–34
low payment of and investment in soldiers by, 21–23, 32, 139
paramountcy of the military during, 2–4, 10–11, 20–22, 194–195
and military fiscalism, 7–9, 84–85

Company raj – *continued*
See also colonialism; East India Company; 1857 uprising; Indian Contagious Diseases Act (Act XIV of 1868); Presidencies; soldiers in the Service of the East India Company
concubines:
and co-habitation, 39
identification as prostitutes, 18–19, 38
See also courtesans; women
contagious diseases:
'natural'/'chronic' diseases and 'preventable' diseases distinguished, 53
See also cholera; disease transmission; gonorrhoea; syphilis; venereal disease
contagious diseases acts,
in Britain, 5–7, 65, 202–203n10
Indian Contagious Diseases Act (Act XIV of 1868), 10, 170, 185–186, 189, 204n32, 216n39
perceived benefit to the military of, 5–7, 131–132, 189
See also Cantonment Act (Act XXII of 1864)
Corwallis' reforms, 30
courtesans:
and co-habitation in pre-colonial India, 39–40
colonial-state influence on the re-definition of, 12, 13, 18–19, 39, 46–47
courts martial:
and 'improper conduct', 153–155
of officers, 147–149, 152, 153–154
rank-and-file cases, 146–149, *150–151t4.3*, 152
and Regulation XX, 132–133
See also punishment
Crimean war, 10, 15, 124, 159, 179
Crown raj:
assumption of power in 1858, 7, 139
continuity with Company raj, 7
paramountcy of the military during, 194
and military fiscalism, 7–8
Cuvier, Georges, *Regne Animal*, 99, 100
daktars (Indians trained in Western medical practices), 93–94
Dalrymple, William, 214–214n11
dancing girls:
devadasi, 40, 205n40, 212n114
prestige and religious function of, 40
re-definition as prostitutes, 18–19, 40, 46–47
targeting for medical examination of, 43, 172
See also nautchees/nautch girls; prostitutes; women
Darwin, Charles, 95–96
debauchery, 2, 4–5, 19, 96–97, 126–127, 182–184
devadasi (temple dancers), 40, 205n40, 212n114
Dirks, Nicholas, 132
disease transmission:
'common sense' on venereal disease control, 11, 14, 86, 89–91, 109–110, 193
Indian women framed as vectors of disease, 13, 50–51, 53–54, 56, 166–167
'natural'/'chronic' diseases and 'preventable' diseases distinguished, 53
soldiers and Company policies, 2–4, 8–9, 49–50, 89, 118, 167
See also cholera
dispensaries, 239n56:
military surgeons' appropriation of, 10–11, 15, 93–94, 104–105, 158, 163, 168, 171–174, 176
the Dutch, 19, 36, 124
dysentery, 48, 53, 56, 126, 162, 181, 190

East India Company:
and the early spice trade, 19–21
limitations on the number of European women in India imposed by, 24–25
marriage of non-Christian women encouraged by, 24
See also Company raj; soldiers in the Service of the East India Company

education, 22–23, 58–60, 124, 137–139, 155–156, 202n5
  See also colleges
Edwardes, S.M., *Crime in India*, 16
1857 rebellion, 124, 130, 158, 159, 179, 191
  regulatory system re-established in the wake of, 10, 15, 183
Esdaile, James, 174
Eurasian children, 'orphan' as a euphemism for, 26
European medical practices, 5, 62–63, 98–99
European troops:
  batta (field pay), 20, 141
  intellectual and social needs of, 22–23, 58–60, 137–138, 202n5
  mortality rates, 7, 91, 202n5
    compared with peers in Europe, 3
  rank-and-file soldiers
    cost of, 2–4, 60, 190–191
    moral improvement, 121, 138
    stereotypes of, 1–4, 59–60, 118–119, 137–139, 148, 148, 156, 190
  recruitment
    and alcohol, 21–22
    molding of recruits, 127–128
    quality of recruits, 3, 16, 137–138, 152
  savings banks for, 60, 138, 202n5
  See also sepoys; soldiers; soldiers in the Service of the East India Company
European women:
  and disorderly conduct, 119, 140–146, *144t4.1*, 156, 176
  limitations on numbers in India of, 24–25
  weak constitutions of, 25, 140

Finch, Assistant Surgeon, 108–109
Fort St George (Madras), 20
Fort William (Bengal), 20
Fort William College (Calcutta), 85, 87
Foucault, Michel, 125
France:
  'continental' or French model for regulated prostitution, 6, 54–55, 181
  liver problems and the French diet, 124
  sepoy battalion raised by French Governor-General Joseph François Dupleix, 21

gender:
  and the framing of the fault for venereal infection, 112–113, 118–119, 192
  'manly', masculine 'urges' as essential to military effectiveness, 2, 8–9, 119
  punitive control over 'unprotected' European women, 141–146
  See also human difference; masculinity; women
Gilbert, Pamela, 94
Goddard, Thomas (Commander-in-Chief of forces at Bombay), 57
gonorrhoea, 3, 10, 18, 38, 51–53, 111, 174
  See also venereal disease
Green, Lieutenant Colonel, 79

Haileybury, 85–88, 96, 221n13
  See also colleges
hakim (practitioner of unani-tibb), 52–53, 108
  See also unani-tibb
Halhed, Nathaniel, *1776 Code of Gentoo Laws*, 37, 52
Harrison, Mark, 96
Hata, Sahachiro, 52
Havelock, Henry (Lieutenant Colonel and Deputy Adjutant General of Bombay), 178, 230n51
  Bible distribution to soldiers organised by, 230n51
  expulsion of 'dissolute' European women by, 142
  rationing of alcohol proposed by, 129
Hawes, Christopher, 30
health:
  of European troops, 2–4, 9, 12–13, 17–18, 55–56, 83
  and managed segregation

health – *continued*
between Europeans and Indians, 85–86, 138, 166, 193
and religious concerns, 74
Hickson, Samuel (officer in the Company army), 25–26, 57, 208–208n44–45
Hodgson, Brian Houghton, 101–102
Hoffman, Erich, 52
homosexuality, 17, 54, 124
human difference:
and European scientific theory, 55, 85–86, 95–103, 126–127, 225n81
Hodgson's varieties and types of peoples, 101–102
morality and 'civilisation' as determining factors, 34–36, 53, 101–105
and the rise of professional societies, 46, 85, 95, 98–99, 110–113, 114, 193
and the treatment of venereal disease in India, 105–108, *107t3.1*, 114
See also class; gender; masculinity; race

Indian Contagious Diseases Act (Act XIV of 1868), 10, 170, 181, 185–186, 186f5.1, 187t5.5, 188f5.2, 189, 204n32, 215n39
See also contagious diseases acts
Indian Penal Code of 1860, 19, 47, 206n5
Indian women:
changing European views of, 19
framed as vectors of disease transmission, 13, 50–51, 53–54, 56, 166–167, 191
relationships with European men, 23–24
indigenous medicine:
and Company writers in the 1630s, 214–214n11
European interest in, 5, 52–53, 99
European reliance on, 62–63, 222n28
and mercury, 108
venereal disease treatments investigated by European surgeons, 105–110
See also ayurveda and unani-tibb
intemperance:
and the canteen system, 129
'healthier' drinks as alternatives, 22–23, 128–129, 138, 156
officers associated with, 123–124, 147
and rank-and-file soldiers, 2–3, 4–5, 58, 118–119, 127–129, 138–139
punishment of, 146–147, *148t4.2*, 149, *150–151t4.3*, 152
temperance societies, 129, 137, 138–139
venereal disease linked to, 120–121
See also liquor

Johnson, James:
career of, 224–224n61
on the effect of the Indian climate on European constitutions, 96–97, 123, 126–127
Jones, William, 52, 98
journals:
Calcutta Medical and Physical Society Quarterly Journal, 103–104, 110, 111
Gleanings in Science, 99
Indian Journal of Medical Science, 100
Journal of the Asiatic Society, 101
Madras Quarterly Medical Journal, 110, 111
Transactions of the Medical and Physical Society of Bombay, 120, 137–138
Transactions of the Medical and Physical Society of Calcutta, 99, 103–104, 110
See also medical and scientific societies

Kipling, Rudyard, *Barrack Room Ballads*, 1, 202n2
Korieh, Chima, 230n54
kotwal (superintendent of police), 57–58, 61–62, 76–77, 219n101

lal bazaars, 2, 9, 13, 24, 26, 49–50, 56–58, 60, 80–81, 117p3, 129–130, 163, 168, 173–174, 208n34
and surveillance, 134–135
by peons, 61–62
kotwal (superintendent of police), 57–58, 61–62, 76–77, 219n101
and the matron system, 61–62
See also prostitutes
Leslie, Matthew (civil servant), 28
Levine, Philippa, 6, 14, 159, 168
liberalism:
and the 'Age of Reform', 84–85, 125
and the closure of lock hospitals, 82–83, 92
concern for soldiers' intellectual or emotional well-being, 60, 85
See also Anglicists
liquor:
Company profits
from canteen sales of, 129
liquor tarriffs and duties on, 230n54
illegitimate selling within the cantonments of, 119, 125, 141
as a noxious and accessible commodity in India, 120
See also intemperance
lock hospitals:
admissions to, 58, *59t2.1*, *67t2.5*, 82, 165–166, *165t5.2*, 183–184, 217n60
Burke's report urging the closure of, 160–165, *163t5.1*
rejection of findings of, 113, 167, 178–179
closure of, 70, 75, 78–79, 82–83, 92, 162–163, 189
cost of, *63t2.2*, *66t2.4*, 82–83, 92, 189
in Europe, 10, 54, 181
'experimental' lock hospital established in Bengal, 10
as hospital-cum-prison, 49, 74–75, 78
and the *lal bazaar* system, 13, 56–58, 81, 163, 168
lock hospital system, 80–81, 160, 171–176, 189, 193–195
and the subversion of state authority, 189
and matrons, 57, 61–62, *63t2.2*, 76–77, 162, 168, 171–173
as military tools, 70–72
proposed for Calcutta, 130
regulation of prostitution in India as a focus of, 77–80, 181, 186, *186f5.1*, *187t5.5*, *188f5.2*
re-instatement of, 14–15, 83, 86, 158, 162, 179, 180, 189
and Murray's report, 113, 164–165, 171, 179
and religious and caste concerns, 74
and the role of native practitioners and indigenous medicines, 62
London Missionary Society, 36, 44–45

Macaulay, Thomas, 93, 168
MacGaveston, Assistant Surgeon, 108
MacLeod, D (Bombay's Deputy Inspector General of Hospitals), 162, 163, 182–183
Macpherson, Hugh (surgeon in the Bengal army), 137
Madras:
Governor Thomas Munro, 32–33, 141
hospital admissions for venereal disease in, *59t2.1*
Madras Military Board and the abolition of lock hospitals, 70, 75
'Mull' (soldier in army of), 21
Regulation VII, 134
See also Fort St George; Presidencies
Madras Quarterly Medical Journal, 110, 111
marriage:
and conversion to Christianity by Indian women, 25
good health associated with domestic life, 182
of non-Christian women encouraged by the East India Company, 24
of orphans to Europeans, 26–27, *27t1.1*
polygamous relationships, 28, 95

marriage – *continued*
  and unofficial Indian companions, 28–36
  See also co-habitation and companionship; courtesans; marriage; prostitutes; women
Martin, James Ranald (Presidency Surgeon of Bengal/President of the East India Company Medical Board), 102–103, 180, 226n98
masculinity:
  bonding/*esprit de corps* amongst European surgeons in India, 14, 86, 193
  bonding of men during college years, 85, 89
  and clubbability, 99
  and the construction of power in colonial India, 154, 156
  manly 'urges' of soldiers, 9, 58, 58
  systematic emasculation of Indian men, 99–103
  See also gender; human difference
masturbation, 17, 54
matrons (*chowdranee/chaudhryan*) and the matron system:
  and corruption, 76–77, 162, 166, 168
  cost/payment of, 162, 172, 173
  and dispensaries, 171–172
  and lock hospitals, 76–77, 162, 168
medical boards:
  'common sense' on venereal disease control, 14, 86, 109–110, 193
  lock hospitals
    deemed too expensive by, 82–83
    sanctioning of, 60
  and the sharing of European knowledge of the subcontinent, 14
  women with venereal disease targeted by, 60, 68, 71
  Bengal Medical Board, 68, 71, 74–75, 82–83, 105–110, 228n4
  Calcutta Medical Board, 172
  Madras Medical Board, 25, 82, 127–128, 139, 172
medical colleges, 89–90, 93–94
medical and scientific societies:
  Bombay Medical and Physical Society, 98, 99, 120
  Calcutta Medical and Physical Society, 98, 103–104, 111
  Madras Medical and Physical Society, 99
  and 'unique' knowledge, 46, 85, 95, 98–99, 110–113, 114, 193
  See also journals; societies
medical topographical reports, 126–127
mercury:
  and different treament methods for Indian versus European bodies, 106, *107t3.1*
  and indigenous medical practices, 108
  Queries relative to the nature and treatment of the veneral disease in India (Appendix 2), 198–201
  side effects of, 106, 108
  venereal disease treatment with, 48, 52, 56, 62, 106–110, 198, 201
military fiscalism:
  and the Company's fixation on the health of European troops, 8–9, 84–85
  and decision making of both the Company and Crown in India, 7–8, 84
military lock hospitals. *See* lock hospitals
missionaries:
  London Missionary Society, 36, 44–45
  morally corrupting influences of India stressed by, 15, 34–36, 44–45
morality:
  and European understandings of 'prostitution', 15, 46–47, 114
  and human difference, 53
    moral character of Eurasian children, 36
  and missionary fund-raising efforts, 34–36
  and mixed marriages, 34–37
  mixed race children associated with, 36
  temperance societies, 138–139

Moseley, Benjamin, 96
Mouat, J. (Surgeon of HM 13th Dragoons at Bangalore), 166–168
Munro, Thomas (Governor of Madras), 32–33, 141
Murray, John (Deputy Inspector General of Hospitals), report supporting lock hospital re-establishment by, 113, 164–166, 171, 179

nautchees/nautch, 12, 38–39, 41–45, 115p1

Omissi, David, 202n4
opium, 62, 109, 149, 228n2
Orientalists/and Orientalism, 84, 96, 220n1
  and the flattening approach to certain Indian roles and traditions, 18–19
orphans:
  marriage to Europeans, 26–27, *27t1.1*
  Military Orphan Society, 26, 33

Pamphleteer, 101
Parkes, Fanny, *Wanderings of a Pilgrim, in Search of the Picturesque*, 29
peons:
  and corruption, 73–74, 77, 166–168
  cost of, 68, 78, 172, 173
  and the surveillance of prostitutes, 61–62, 69, 74, 76, 79, 162–163, 173
phrenology:
  Calcutta Phrenological Society, 101
Piddington, Henry, 128–129
the Portuguese, 24, 29, 34, 36, 124
Presidencies:
  armies/military of
    formation of, 20–21
    Indian Army formation from the merging of, 20, 206–207n10
    paramountcy of, 20–22, 194–195
    rivalries between, 21
    role of regulation in the construction of military space, 136
  medical boards in presidency capitals, 90, 94, 164
  presidency capitals in Bengal, Madras, and Bombay, 20
    comparative state of venereal diseases in, *165t5.2*, 183–184
    and the India Contagious Diseases Act, 204n32
    and the lock hospital system, 183–186
  regional differences
    and the Cantonment Act (Act XXII of 1864), 184
    elevated status of Bengal, 20
    and sepoy families, 241n110
  See also Bengal; Bombay; Calcutta; Madras
Price, Surgeon (HM 12th Foot at Trichinopoly), 56, 58, 69
Prichard, James Cowles, 95–96, 99–100, 225n81
prostitutes and prostitution:
  classification of sexual companions as, 56–57
  'continental' or French model for regulated prostitution, 6, 54–55
  as 'criminals', 19, 47, 58, 77–78, 81–82, 206n5
  demonisation of, 113–114, 191–192
  and European moral categories, 12, 15, 18–19, 38–40, 46–47
  European women as, 140–142, 143, *144t4.1*, 156
  Indian and Eurasian wives accused of devoting their young children to prostitution, 36
  'low-class' females associated with venereal disease transmission, 53, 184–185, 205n44
  regulation/policing of
    by matrons, 61–62, 76–77, 162, 171–173
    and corruption, 77, 162, 166–168
    and the Indian Contagious Diseases Act (Act XIV of 1868), 185–186, *186f5.1*, *187t5.5*, *188f5.2*
    in the *lal bazaar*, 2, 4, 12, 24, 49–50, 56–58, 60

prostitutes and prostitution – *continued*
punitive methods of control, 5–6, 159, 166
unofficial census taken of, 176, *177t5.4*, 178
as security risks, 4–5, 184–185
See also Cantonment Act; *lal bazaar*; lock hospitals
public health in India:
and dispensaries, 171, 174
and the health of European soldiery, 10–11, 71–72
and the lock hospital system, 174
punch houses, 129–130
punishment:
corporal punishment of Indians selling unlicensed liquor in the cantonment, 134
courts martial, for sexual impropriety, 153–155
and disorderly conduct of European women in the cantonment, 119, 140–146, *144t4.1*, 156
of European troops
corporal punishment, 122, 146, 148–149
infliction of, 146–147, *148t4.2*, 149, *150–151t4.3*
of European troops with venereal disease, 3–4, 9, 17–18, 55–56, 83
Indian Penal Code of 1860, 19, 47, 206n5
lock hospitals as hospital-cum-prison, 49, 74–75, 78
of "unnatural" acts, 153–155
of women deemed to be prostitutes, 5–6, 159, 166
See also Cantonment Act (Act XXII of 1864); courts martial

Qui Hi, 1, 21, 117p3, 202n1

race:
certificates of descent and educational establishment, 33–34, *35t1.2*
and European troops, 2, 137, 190
European women as symbols of white womanhood, 141–146
and medically driven 'morality', 11–12, 14, 24
Eurasian children
exclusion from political and military office, 30
moral failings associated with, 36
mixed-race relationships, 30, 32–37
See also human difference
Regulating Act of 1773, 20, 206n7, 210n94
regulations:
Regulation III, 132–133, 229n24
Regulation VII, 134
Regulation XX, 132–133
Report of the Commissioners Appointed to Inquire into the Sanitary State of the Army in India (1863), 159, 181–182, 213n1
Roman Catholics/Portuguese Catholics, 24, 29
Royal Commission on the Sanitary State of the Army in India (1859), 159, 180–185, 226n98

Schaudinn, Fritz, 52
Scott, James, 49, 83
sepoys (Indian soldiers in the employ of European forces):
'character' of, 22, 103–104
cost of, pay in the 1830s of, 8
defined, *207n12*
families/children of, 64, 92, 241n110
health of, 22, 64–65, 72, 93, 222n28
literacy of, 202n4
numbers of, 2–3, 7, 21–22, 46, 204n24, 208n44
rebellion in 1857 of, 191
See also soldiers
sexual impropriety, punishment of "unnatural" crimes, 153–155
Sinha, Mrinalini, 99, 154
Skey, Frederic Carpenter, 112
slaves and servants:
and female companions, 28–29
and Regulation III, 132–133, 224n24
Smith, George (Assistant Surgeon), 109
Smith, Major Charles H, 4, 99

societies:
  temperance societies, 129, 137, 138–139
  See also medical and scientific societies
soldiers:
  and Company policies, 2–4, 8–9, 49–50, 89, 118, 167
  behaviour of, 2, 8–9, 119, 148, 156
  marriage of, 26–27, *27t1.1*
  See also European troops; sepoys
soldiers in the Service of the East India Company:
  Sergeant John Pearman, 22, 27, 32
  Sergeant Jonathan Taylor, 235n130, 236n147
Stanley, Peter, 16, 228n1
Suleri, Sara, 42, 124–125
syphilis, 108, 111:
  control measures, 3, 10, 18, 38
  identification of *treponema pallidum*, 46, 52
  Salvarsan as a cure for syphilis, 52, 159
  threat of compared with cholera, 161
  See also venereal disease

temperance:
  Chevers warnings regarding drunkenness among European troops, 120
  and 'healthier' drinks, 22–23, 128–129, 138, 156
  temperance societies, 129, 137, 138–139
  See also intemperance
Thornton, Robert John, 97
tropical medicine, and 'natural'/'chronic' vs. 'preventable' diseases, 53
Tulloch, Sir Alexander Murray, 180–181

unani-tibb, 14, 52, 93–94
  See also ayurveda; indigenous medicine

vaidyas/vaids (practitioners of ayurvedic medicine), 104
  See also ayurveda

venereal disease:
  as an imperial threat, 97, 103, 157–160, 164, 182, 190–195
  cases among European troops of, 48, 58, *59t2.1*, 66–68, *67t2.5*, *67t2.6*, 82, 105, 112, 121, 157, 165-167, *165t5.2*, *175t5.3*, 183–184, 190
  and hospital admissions, 58, *59t2.1*, *67t2.5*, 82, 172–174, *175t5.3*, 180, 190–191
  as a 'preventable' disease, 53–54, 159, 181
  Salvarsan as a cure for syphilis, 52, 159
  and sepoys, 64–65, *64t2.3*
  transmission, 2–4, 8–9, 51–54, 89, 108, 118, 167
  See also gonorrhoea; syphilis

Weatherall, GA (Secretary to the Commander in Chief, Fort Saint George), 141
Wellesley, Arthur, 22, 122, 124, 139
Williams, Monier (Surveyor-General in Bombay):
  on drunkenness among English soldiery, 123, 147
    and canteen policies, 128
Williamson, Captain Thomas (former officer in the Bengal army), 209n62
  East India Vade Mecum, 29
women:
  certificates of descent and educational establishment, 33–34, *35t1.2*
  classified as prostitutes, 56–57
  the cost of Indian wives and mistresses
    compared with British wives, 29
    and Company profits, 2–3
  Portuguese wives, 24, 29
  See also concubines; courtesans; dancing girls; European women; gender; Indian women; marriage; *nautchees/nautch* girls; orphans

Printed and bound by CPI Group (UK) Ltd, Croydon, CR0 4YY